Sustainable Materials: The Role of Artificial Intelligence and Machine Learning

Editors

Akshansh Mishra
Politecnico Di Milano, Milan, Italy

Vijaykumar S. Jatti
Symbiosis Institute of Technology, Pune, India

Shivangi Paliwal
Department of Mechanical Engineering
University of Kentucky, Lexington, KY, USA

CRC Press is an imprint of the
Taylor & Francis Group, an **informa** business

Cover credit: by [Cambridge Spark] via Pixabay

First edition published 2025
by CRC Press
2385 NW Executive Center Drive, Suite 320, Boca Raton FL 33431

and by CRC Press
4 Park Square, Milton Park, Abingdon, Oxon, OX14 4RN

CRC Press is an imprint of Taylor & Francis Group, LLC

Library of Congress Cataloging-in-Publication Data (applied for)

ISBN: 978-1-032-56852-2 (hbk)
ISBN: 978-1-032-56853-9 (pbk)
ISBN: 978-1-003-43736-9 (ebk)

DOI: 10.1201/9781003437369

Typeset in Times New Roman
by Prime Publishing Services

Preface

The superior multi-functional properties of polymer composites have made them a suitable candidate for biomedical, defense, automobile, agriculture, and domestic applications. The growing demand for these composites calls for an extensive investigation of their physical, chemical, and mechanical behaviour under different exposure conditions. Characterization techniques are very vital considering the extensive investigation. The criticality and consideration of a number of parameters for the characterization make this investigation more complex. The self-learning ability of machine learning algorithms makes this investigation more accurate and accommodates all the complex requirements. The recent development in neural codes can accommodate the data in all the forms such as numerical values as well as images.

The aim of this edited book is to address the design, characterization, and development of prediction analysis of sustainable polymer composites using machine learning algorithms. Recent development in AI & ML techniques help in the development of sustainable development polymers. The development of AI & ML technique for effective characterization based on the research data available for different polymers. Considering the capability of AI&ML techniques, various properties such as physical, mechanical, chemical, thermal, and electrical can be predicted for desired sustainable polymer composite.

This consists of eleven chapters. The methodology, applications, and consequences of AI and ML in material science are explained in Chapter 1, along with the problems and opportunities that lie ahead. It is a potent union of human experience and AI-driven insights that will revolutionize material creation and design across many industries.

Chapter 2 focused on the structural stress distribution in ASTM D3039 tensile specimens of carbon-epoxy and kevlar-epoxy based composite materials is determined using a data-driven artificial intelligence approach.

Chapter 3 presents detailed characterization of the relationships between composition, processing, and microstructure may be one advantage of quantitative image analysis. The study demonstrates how optimized compositions and processing parameters might affect the microstructure to yield perfect and enhanced segmentable MMCs.

Chapter 4 demonstrates the environment and sustainable concrete. Cane sugar Bagasse ash is already seriously polluting the environment, so disposing of the garbage urgently is necessary. Thus use of wastes like cement and fine aggregate replacement materials can lower the cost of producing concrete and lessen the negative environmental effects associated with disposing of these wastes.

Chapter 5 focus on developing a computational material or cheminformatics feature description language for the protection of valuable assets such as sculptures, priceless decorations, and rare stones in huge quantities of Jewels.

Chpater 6 explains the car explicit dynamic crash analysis with metal, composite, and alloy materials.

Chapter 7 depicts the JAYA and Cohort Intelligence Algorithm for Friction Stir Spot Welded ABS Weld Strength Optimization.

Chpater 8 presents the classification of FDM 3D printed samples' dimensional deviation using supervised machine learning.

Chapter 9 demonstrates the flexural strength estimation of polymer composites via the K-nearest neighboring classification algorithm.

Chapter 10 explains the method of classifying the surface roughness of fused deposition modeling 3D printed samples via supervised machine learning.

Chapter 11 presents impact strength estimation of polymer composites using the K-nearest neighboring classification method.

Akshansh Mishra
Vijaykumar S. Jatti
Shivangi Paliwal

Contents

CHAPTER

1

Artificial Intelligence in Material Science

Sarvat Zafar[1*], Nadim Rana[2]

[1] Department of Chemistry, Faculty of Science, Jazan University, Jazan, Saudi Arabia

[2] Department of IT and Security, College of CS and IT, Jazan University, Jazan, Saudi Arabia

1. Introduction

Exploring novel materials will play a crucial role in advancing human development. After centuries of dedicated research and development, the field of materials science has accumulated a vast amount of knowledge. However, due to our limited cognitive capacity, it can be challenging for humans to keep pace with the constant influx of new information. Only a fraction of the available data in any area can be thoroughly examined. Present-day material studies heavily rely on the "trial-and-error" approach, involving numerous experiments guided by experts, complemented by a limited number of computer simulations. This method demands significant investments of time, money, and resources. Many valuable datasets still need to be discovered or utilized in the archives [1]. Traditional techniques are constrained due to technological advancements in material assessment and the rapid expansion of extensive and multidimensional data. Despite the abundance of information available, the sheer volume of data poses a challenge for experts regarding comprehension and effective management [2]. Only a fraction of this data can be thoroughly examined within specific fields. Scientists often adopt an iterative approach, conducting experiments based on existing knowledge, occasionally with the assistance of computer simulations. This process demands substantial resources and time, leading to slow progress [3]. Consequently, a substantial amount of material-related information requires sluggish database updates and needs to be utilized more efficiently. This

*Corresponding author: sarvatzafar@gmail.com

emphasizes the pressing need for new research methods to accelerate material innovation. Therefore, it is imperative to explore new research methodologies to accelerate the development of innovative materials. The emergence of AI represents a new frontier in advancing the physical sciences field [4, 5].

The emergence of AI also brings intriguing prospects for material science [6, 7]. Over the past six decades, AI has progressed from elementary models to intricate neural networks, laying a solid foundation for its application in material science. The data analysis capabilities of AI have reunited awareness from the material science community [8, 9]. This advancement has established a robust algorithmic framework and dependable hardware support for AI within material science [10–12]. Kalidindi et al. [13] introduced the concept of leveraging big data and AI to propel scientific innovation, particularly in addressing the intricate challenges encountered in material research. This concept gave birth to materials informatics, a consolidation of materials science and AI techniques. Materials informatics aids scientists in unveiling screened connections among various factors, predicting material properties, optimizing processes, and refining material analysis methods. ML [14, 15], a significant feature of AI, is rapidly evolving and holds immense potential for material science. The field of materials informatics, which blends materials science with AI techniques, emerges as an interdisciplinary domain assisting scientists in revealing screened connections among diverse factors. This aids in predicting specific material properties, guiding material development via chemical synthesis, optimizing process parameters, and enhancing established methods for material analysis [16–18].

Machine learning, a substantial component of artificial intelligence, is making rapid advancements and is an exciting pathway for AI's contributions to the study of material science. Over the past decade, machine learning has experienced significant growth and is now making its way into scientific disciplines, including material science. Machine learning aids in characterizing materials, predicting properties at the molecular level, expediting simulations, discovering new materials, and constructing models for designing novel materials [18, 19].

2. Artificial Intelligence Techniques in Material Science

In this section, we take an in-depth look at the utilization of AI techniques, their practical applications, and the resulting implications for researchers in the field of material science. We explore how these AI techniques are effectively harnessed to address various challenges and opportunities within material science.

2.1 AI Datasets in Material Science

Datasets are essential for AI models, including those used in material science and engineering. We require datasets with specific properties and sufficient examples to use AI in this field effectively. The National Institute of Standards and

Technology has recently produced valuable datasets for material engineering. One such dataset, JARVIS-DFT, contains properties like formation energies, bandgaps, and more for 40,000 bulk and 1,000 crystalline materials. As illustrated in Fig. 1, high-throughput reaction screening is used in scientific and industrial research to efficiently evaluate many chemical reactions simultaneously. The Density Functional Theory descriptor calculation is employed to gain intricate electronic and structural insights into these reactions, aiding in predictive analysis. Machine learning, notably the Random Forest algorithm, is applied to data generated during screening to discern critical variables. Effective data input management is integral for organizing and optimizing this process, enhancing research and development efficiency. In 2017, Ward and coworkers [20] conducted a study where they used machine learning models, specifically decision tree models, to link the calculated energy needed to form a material with specific characteristics obtained from the structure of crystals. The models they built using this approach could predict formation enthalpies with an average error of 80 milli-electron volts per atom during cross-validation. Agrawal et al. [21] worked on creating and making available predictive models for the energy required to form materials. They used data from nearly 100,000 calculations based on a method called Density Functional Theory (DFT). These models can be accessed through an online tool to estimate formation energy. In a study, Choudhary et al. [22] research focused on finding materials that can efficiently convert heat into electricity, known as thermoelectric materials. They used physics-based transport methods, DFT calculations, and machine learning models to discover these materials through data analysis.

The Wolverton Research Group at Northwestern University manages an important dataset called the Open Quantum Material Database. This dataset offers valuable details about the thermodynamic and structural properties of a substantial collection of 1,022,603 materials. Kirklin [23] assessed the accuracy of density functional theory (DFT) formation energies in the OQMD and found that the database contains over 300,000 calculations detailing the electronic structure and stability of inorganic materials. Shen et al. [24] reflected on the growth of the OQMD, which now includes over one million compounds and is constantly used by researchers worldwide. Saal and their team [25] highlighted the power of high-throughput DFT databases like the OQMD in accelerating materials design and discovery, with over 200,000 DFT-calculated crystal structures available.

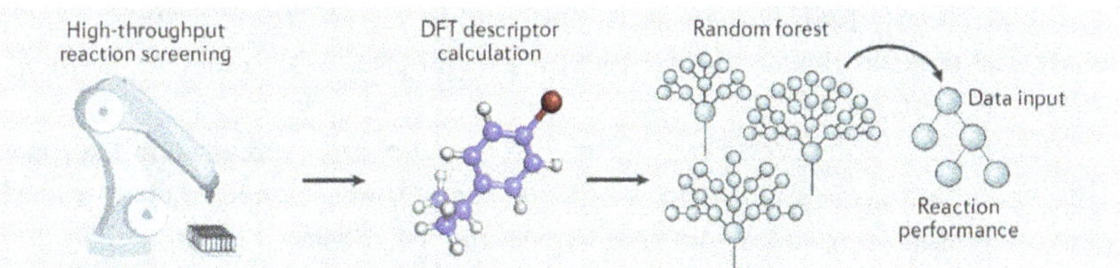

Figure 1: Optimizing ML model development with high-throughput reaction screening data [26]

2.2 Neural Networks (NN) in Material Science

Neural networks are like computer systems inspired by the human brain. They have three main parts: input, hidden, and output layers. The input layer receives data, the output layer provides the final result, and the hidden layer(s) perform complex calculations and find patterns. Each layer contains artificial neurons that connect to neurons in other layers through weights. These weights determine the strength of connections and are adjusted as the network learns from data. A neuron's role is to add inputs, considering these connection strengths, and then apply a specific function to the result. This function is typically non-linear, allowing the network to understand intricate relationships in the data. So, a standard neural network is structured, as shown in Fig. 2.

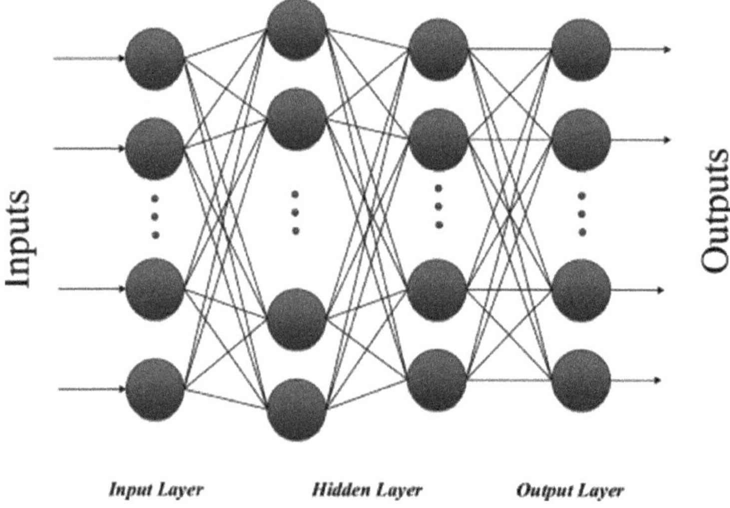

Figure 2: Neural networks representation in machine learning [27]

Ihom et al. [28] emphasizes that neural networks are a powerful predictive tool for solving complex problems in this field. It also compares neural networks to linear regression models, finding that neural networks are more sophisticated in providing solutions. In the research conducted by Segler [29] and his team, they employed three unique neural networks to act as strategies for Monte Carlo Tree Search (MCTS), as illustrated in Fig. 3. MCTS is a computational method for making decisions and planning strategies, which can be applied in various scenarios.

Zhang and their coworkers [30] discuss the widespread application of artificial neural networks in various aspects of materials science, such as material design, preparation, processing, and corrosion. Bhadeshia [31] and their team emphasize the importance of neural networks in addressing complex problems in materials science and engineering, providing longer-term solutions. Finally,

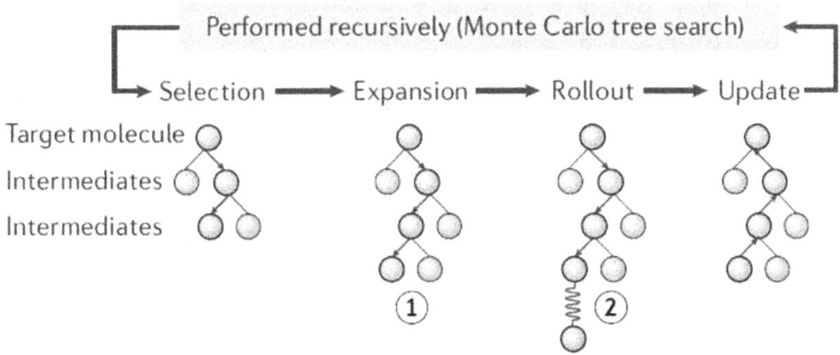

Figure 3: Chemical synthesis strategy planning using Monte Carlo Tree Search [29]

Dobrzanski [32] presents the broad possibilities offered by artificial intelligence tools, particularly neural networks, in material engineering, including modeling, characterization, and forecasting.

In material science, several specialized types of neural networks are commonly employed. These include Convolutional Neural Networks (CNNs), Graph Neural Networks (GNNs), and Generative Models. In a Convolutional Neural Network (CNN), a crucial mathematical operation called convolution plays a pivotal role [33]. This operation blends two functions to generate a third function that illustrates how one of the original functions influences the other. In the context of a CNN, it involves taking input data, often in the form of an image, and passing it through a set of filters that can adapt and learn. These filters yield what we refer to as "feature maps". In a CNN, the input data typically takes on a 3D structure with dimensions such as width, height, and channels. The channels represent different aspects of the data, such as the color channels in an image (e.g., three channels for RGB images). Conversely, the filters, also known as kernels, also exist in a 3D structure with dimensions matching those of the input channels. To apply these kernels to the input data, a convolutional layer within a neural network is employed. In this process, the kernel moves across the input data. It calculates the dot product at each position with a local section of the input data. This calculation results in a single value, which is then positioned in the corresponding spot of the output feature map.

Mathematically expressing this convolution operation, we represent it as $G = F * K$, where F denotes a matrix representation of the input data, K signifies a matrix representation of the kernel, and G denotes a matrix representation of the resulting output feature map [10]. In research by Sorini and colleagues [34], they introduce a Convolutional Neural Network designed to mimic the stiffness of composite materials precisely. They aim to use this CNN to forecast how damage progresses in composite structures. On the other hand, Kalliatakis et al. [35] assess different CNN designs to classify materials. Their work achieves a high level of accuracy in distinguishing and classifying materials based on images.

2.3 Enhancing Material Science Efficiency through Graph Neural Networks

A Graph Neural Network (GNN) is a computational system that processes data in a graph format. GNNs have gained popularity due to their proficiency in handling intricate connections and relationships among entities within a graph. In a graph, entities are referred to as nodes (or vertices), and these nodes are connected by lines called edges. Each node and edge can possess specific attributes, such as labels, numerical values, or categories. GNNs function by inspecting a node and the nodes and edges directly linked to it. They employ a specialized mathematical model called a neural network to gather information from these neighboring elements and adjust and enhance their node representation. This process can be iteratively repeated, allowing the network to learn progressively more complex patterns and connections within the data. GNNs have proven highly effective across diverse domains, such as recommendation systems, drug discovery, social network analysis, and traffic prediction. These networks are versatile and can work with various types of graphs, whether directed or undirected, involve different entities, or exhibit a wide range of characteristics [36].

Ronchetti and co-authors [37] apply GNNs to predict physical properties and optimize cathode materials for batteries, demonstrating the effectiveness of GNNs in accelerating materials discovery. Omee et al. [38] propose a scalable and deeper GNN model, DeeperGATGNN, which achieves state-of-the-art performance in predicting crystal structure energy and band gap properties. Allotey and their colleagues, as mentioned in reference [39], have presented an innovative approach to active learning using entropy. They combine GNNs with Gaussian processes to predict the properties of materials and pinpoint regions in the chemical space where the model lacks confidence. In summary, these research papers emphasize the important role of GNNs in materials science. They show that GNNs can be quite effective in predicting material properties, speeding up the discovery of new materials, handling large datasets, and making intelligent choices about which data to use during the modeling process.

2.4 Generative Models in Material Science

About three decades ago, John Maddox [16] noted that a significant challenge in physical science was the inability to predict the arrangement of atoms in even the simplest crystalline materials based solely on their chemical composition. Since then, various methods have been developed for predicting the structures of molecules and solids. However, they tend to be computationally expensive and require searching through a vast space of possible structures. Deep learning algorithms, specifically Generative Networks (GN), have been introduced to address this challenge in recent years. GNs can generate potential structures based on given input information. In simpler terms, it provides the model with specific details and can provide a range of possible structures as its output. This represents a promising approach to tackle the long-standing problem of crystal structure prediction more efficiently and effectively. In the research conducted by Harshvardhan [40], a comprehensive review and analysis of various generative models are provided. This research investigates specific models, namely Gaussian Mixture Models, Hidden Markov Models, and Generative Adversarial Networks. The study explores how these models function and can be practically applied in various contexts.

3. Integrating AI into Materials Science

In recent years, AI has made substantial progress in various fields, including materials science. Machine learning, in particular, has gained recognition for its ability to assist in developing new materials and predicting chemical combinations for the synthesis process [41, 42]. In the next section, we will explore how ML is positioned to tackle the challenges related to material design, creation, and management, effectively bridging these aspects [43–45].

3.1 Enhancing Computational Speed

The research process in computational chemistry and materials science has evolved into three generations. The first generation used local optimization algorithms to predict material performance based on structure. The second generation focused on predicting structure and performance using global optimization algorithms to forecast the crystal structure based on the elements in the composition. The third generation, known as 'statistically driven design,' now employs Machine Learning algorithms to predict elements' composition, structure, and performance using physical and chemical data [16, 45, 46]. However, there are challenges in discovering high-performance materials due to theoretical limitations. Practical conditions like mixed phases or grain boundaries must match the model parameters. For instance, when utilizing Density Functional Theory, a zirconium-doped lithium tantalum silicate was expected to have a conductivity of 10^{-3} S cm^{-1}, but experiments showed it to be around 10^{-5} S cm^{-1} [47]. Therefore, exploring how Machine Learning can help address these simulation limitations is crucial.

3.2 Atom2Vec: Advancing Materials Discovery

The Atom2Vec program, a tool in unsupervised Machine Learning, accomplished the recreation of the periodic table of elements within a few hours [48, 49]. This program begins by learning to distinguish between different atoms by analyzing compounds within an online database. It employs a concept similar to language understanding, where the attributes of a word are discerned by considering the surrounding words. In the case of Atom2Vec, it clusters chemical elements based on their chemical characteristics. Furthermore, the vector-based atom descriptions it generates contain extensive information about the patterns of elements as they appear on the periodic table. This unique descriptor has the potential to serve as input for various Machine Learning models, offering a practical and innovative method to represent material data in the future quantitatively [49, 50].

3.3 Expanding Simulation Analysis

Machine Learning can identify predictable patterns in theoretical calculations for atomic force fields, making energy and force field calculations faster. This allows for simulating the movement of millions of atoms within a few nanoseconds, even if the initial focus was on a smaller number of atoms and a shorter time frame, such as a few picoseconds. Scaling up in this way significantly extends the duration and scope of simulations, leading to more accurate results. As a result, complex structures like amorphous or polycrystalline materials and chemical reactions like corrosion and interfacial reactions can be simulated [51–54]. Developing precise interatomic potentials is challenging when dealing with large-scale simulations investigating chemical processes at surfaces and interfaces. This complexity arises from the diverse atomic environments and various types of bonds in these systems. A novel approach uses artificial neural networks to establish interatomic potentials. This method offers an unbiased means of describing potential energy surfaces, particularly for systems where traditional potentials struggle to provide accurate representations [55, 56].

Artrith and a team of researchers performed experiments utilizing an artificial neural network-based approach on copper and zinc oxide systems. They employed this technique to explore how copper interacts with zinc oxide on oxide surfaces. The results showed that the neural network's potential energy calculations closely matched those obtained from fundamental electronic structure calculations, but the neural network approach was notably faster. While creating these neural networks requires substantial computing power for extensive training data, their advantages become apparent when dealing with large-scale applications where conventional methods face challenges [57].

3.4 Simplifying Computational Load

The extensive range of material combinations in science makes exploring all possibilities using traditional simulation methods within a reasonable timeframe impractical. However, a more efficient approach is to train an ML model with a subset of available data and then use this model to forecast outcomes for other combinations. This method substantially simplifies computations and significantly speeds up the filtering process [58, 59]. Panapitiya and colleagues introduced an ML model using the stochastic forest method to forecast the CO adsorption energy of nanoclusters. They initially utilized data from DFT simulations of silver (Ag) alloyed gold (Au) nanoclusters to train the model [60]. Through closely examining the critical factors in the random forest model, the researchers identified that the arrangement of Ag atoms within Au_{25} significantly influences CO adsorption sites. This ML model can quickly expand to predict the behaviour of other nanoclusters based on gold. It serves as a valuable screening tool for identifying promising materials for more thorough and accurate analysis in the future.

3.5 Predicting Material Traits via Structure-Property Connections

One exciting aspect of material science involves predicting the properties of new materials by understanding their structural characteristics. Scientists employ advanced techniques and computer models to analyze materials' atomic and molecular structures. They then use this information to create maps or models illustrating the relationship between structure and specific properties. These maps serve as a valuable tool for anticipating the traits of new materials. This approach is valuable because it saves scientists time and resources. Rather than conducting numerous lab experiments, they can make informed predictions about which materials are worth investigating [44, 61, 62]. Material researchers often strive to enhance specific material properties, such as improving electrolyte conductivity, enhancing the Seebeck coefficient in thermoelectric materials, or increasing the efficiency of organic-inorganic hybrid perovskites [63, 64]. They typically rely on trial-and-error experiments, sometimes guided by theoretical simulations or the expertise of chemists. However, these methods can be time-consuming and may only sometimes yield optimal results. Fortunately, Machine Learning models offer a promising solution. These models can reasonably predict material properties and structures before manufacturing begins [44, 65]. For example, Sendek and their team utilized an ML model developed in MATLAB to select a small group of specific solid electrolytes from a pool of over 12,000 materials [48].

ML can rapidly assess various options through experiments and simulations. For instance, Li and their research team used machine learning and high-throughput simulations to investigate the stability of double perovskite halides [66]. They initiated the process by creating a database of decomposition energy, which relies on calculations using DFT, a computational method closely linked to stability, for 354 different perovskite choices. This database was then used to train an ML model. The model's predictions were subsequently confirmed by examining the structure of perovskite in 246 compounds, demonstrating a high level of accuracy (with an F1 score of 95.9%). This study highlights the fact that ML predictions are more efficient and effective than actual experiments. Similar approaches have been applied in the development of lead-free organic-inorganic hybrid perovskites [67, 68], the design of single-atom catalysts [69], the production of light-emitting diodes (LEDs) [70, 71], and the creation of other essential materials [72].

3.6 Material Synthesis Route Design

Scientists have proposed specific techniques for creating computer programs to address fabricated challenges in organic synthesis [29, 73, 74]. To computer scientists, a chemical reaction resembles information that reveals how different compounds link together. This information can be structured and represented using a data format similar to a graph or network. AI then comes into play by interacting with these structures to propose the most efficient synthesis pathways (as shown in Fig. 4) [56–58]. Kim et al. collaborated to utilize ML and natural language processing techniques for extracting synthetic conditions from published literature [59]. They designed an AI platform that automatically analyses literature and categorizes it based on the mentioned keywords, such as synthesis temperature, time, equipment, preparation details, and target materials. This platform achieved a remarkable 99% accuracy in identifying sections and an 86% accuracy in tagging keywords. Using this platform, the team examined synthesis conditions for various metal oxides across 12,900 pieces of literature.

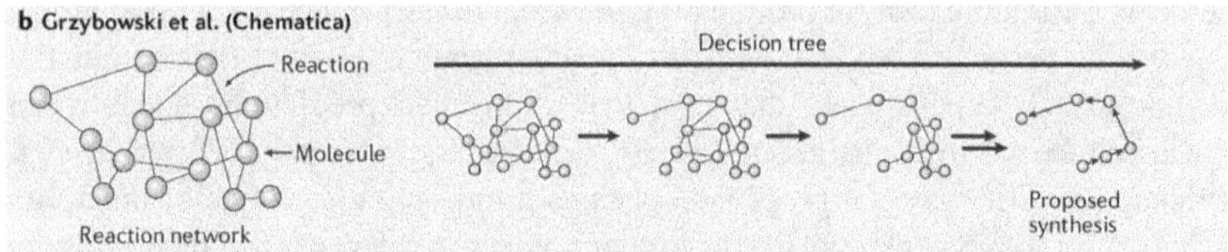

Figure 4: Utilizing decision trees for identifying chemical synthesis procedures in complex reaction networks [75]

Huo and their team [76] developed an ML-based method to gather and organize information about making inorganic materials from many written documents. They used a technique called the latent Dirichlet allocation (LDA) model to group keywords related to specific steps in material synthesis, like "grinding" or "heating". By analyzing over 2.2 million documents, they identified different synthesis methods. With a small amount of labeled data, they trained a computer program to sort these methods into categories like "solid-state" or "sol-gel". Machine learning can effectively organize and explain how materials are made, which could help build a comprehensive materials database. Raccuglia et al. [77] also used ML models, trained on data from experiments that went differently than planned. These models were surprisingly successful, with an 89% accuracy rate in predicting how to make new materials that combine organic and inorganic components. Typically, research papers in chemistry focus on successful experiments, but this study emphasizes the value of learning from unsuccessful ones to better understand the conditions for success.

3.7 Optimizing Experimental Factors

In this area, researchers have actively conducted numerous studies to enhance the performance and applications of AI by optimizing experimental factors in material science. To achieve their desired outcomes, these studies actively explore factors such as material composition, fabrication techniques, testing methods, and data analysis approaches [8, 78]. As a result, many investigations actively assess how these factors impact the efficiency and effectiveness of AI algorithms, ultimately leading to advancements in materials design for artificial intelligence systems. This application of AI actively improves experimental parameters and, in turn, advances the field of material science [79, 80]. The available literature comprehensively explores various aspects of optimizing experimental factors for material science and AI. Numerous studies have focused on specific elements of this endeavour, encompassing a wide range of material types, including metals, polymers, ceramics, and composites. These studies have analyzed the properties and behaviour of these materials when integrated into AI technologies [6, 81, 82]. Researchers have extensively examined the impact of factors such as material characteristics (e.g., electrical conductivity, thermal stability), processing conditions (e.g., temperature, pressure), and data acquisition techniques (e.g., imaging, sensor data) on the overall performance of AI systems [83, 84].

In their study, Zhang and the research team [85] introduced a novel combination of machine-learning techniques to identify optimal operational strategies for the Aerosol Jet 3D Printing (AJP) process in various design scenarios. This approach involves using conventional ML methods such as experimental sampling, organizing data, categorizing it, and leveraging prior knowledge. As material manufacturing processes transition toward greater automation, they become increasingly integrated with advanced industrial manufacturing practices, often called Industry 4.0. For example, programmable systems play a significant role in rapidly producing polymers [86, 87]. Machine learning is pivotal in these situations, especially during the initial phases of implementing these streamlined synthesis processes. It helps precisely determine the quantities of raw materials and catalysts needed to synthesize organic compounds that meet specific criteria, such as achieving the desired weight, maintaining a narrow distribution range, and minimizing undesired side effects [88, 89].

3.8 AI Integration in Traditional Analytical Instruments

In recent years, a notable trend has emerged in analytical instrumentation involving integrating AI techniques into conventional instruments. This blending of AI and traditional instruments shows significant potential for boosting the capabilities of these tools and enhancing their accuracy, efficiency, and depth of information. AI techniques have been gaining prominence in conventional materials science and engineering (MSE) instruments like X-ray diffraction (XRD), Transmission Electron Microscopy (TEM), and Scanning Electron Microscopy (SEM), with a particular focus on image analysis [90–93]. In the case of XRD, conventional methods for analyzing patterns and creating phase diagrams have been time-consuming and labour-intensive. However, recent advancements have significantly streamlined this process by seamlessly integrating ML models. These models can efficiently identify compounds of interest and generate phase diagrams, ultimately saving time and effort [94, 95]. Researchers employed an advanced computer program called a Convolutional Neural Network to precisely align the XRD spectra of materials from an experimental database, effectively eliminating unwanted background noise. CNN outperformed traditional ML methods, achieving an impressive accuracy rate of 96.7% [96]. Furthermore, researchers have developed a computer program called "Auto detect mNP", capable of autonomously identifying the shapes of metal nanoparticles (mNP) in images captured using a Transmission Electron Microscope. Additionally, it can detect impurities in the mNP production process and distinguish between long rod-shaped nanoparticles and shorter ones [97]. Using high-resolution TEM (HRTEM) data, the researchers devised a two-step nanoparticle analysis approach [98, 99]. Firstly, they applied a computer program with a U-net design, CNN, to outline nanoparticles within HRTEM images. Subsequently, they employed a random forest classifier to identify anomalies within specific nanoparticle segments. This system achieved an accuracy rate of approximately 86% [100].

4. Conclusion

Integrating machine learning and artificial intelligence into material science signifies a profound transformation. In contrast to the traditional reliance on expert knowledge and labour-intensive approaches, this field now harnesses the capabilities of ML and AI. Researchers employ ML techniques to expedite material characterization, predict molecular-level properties, accelerate simulations, discover novel materials, and craft innovative designs. Nevertheless, challenges persist due to technological advancements and the data deluge. These challenges underscore the need for novel methodologies to accelerate material innovation—the advent of AI ushers in promising prospects and materials informatics. By fusing materials science with AI techniques, we can unveil concealed correlations, forecast material properties, optimize processes, and enhance material analysis techniques.

On the other hand, machine learning, a pivotal facet of AI, is advancing swiftly and carries substantial potential for material science. This study tries to put forward the methodologies, applications, and ramifications of ML in material science, exploring many possibilities and challenges. The collaboration between human expertise and AI insights is set to revolutionize material design and development across various industries.

References

1. Maguire, J.F., Benedict, M., Woodcock, L.V. and LeClair, S.R. 2001. Artificial intelligence in materials science: Application to molecular and particulate simulations. *MRS Online Proceedings Library (OPL)*, 700: S8.1.
2. Abdalla, H.B. and Abuhaija, B. 2023. Comprehensive analysis of various big data classification techniques: A challenging overview. *Journal of Information & Knowledge Management*, 22(01): 2250083.
3. Zahrt, A.F., Henle, J.J., Rose, B.T., Wang, Y., Darrow, W.T. and Denmark, S.E. 2019.

Prediction of higher-selectivity catalysts by computerdriven workflow and machine learning. *Science*, 363: eaau5631.

4. Moses, K., Michaels, S. and Willis, M. 2020. Physics and devices toward artificial intelligence. *Japanese Journal of Applied Physics*, 59: 060401.
5. Wang, Z., Sun, Z., Yin, H., Liu, X., Wang, J., Zhao, H. et al. 2022. Data-driven materials innovation and applications. *Advanced Materials*, 34(36): 2104113.
6. Huang, J.S., Liew, J.X., Ademiloye, A.S. and Liew, K.M. 2021. Artificial intelligence in materials modeling and design. *Archives of Computational Methods in Engineering*, 28: 3399-3413.
7. Kline, R. 2010. Cybernetics, automata studies, and the Dartmouth conference on artificial intelligence. *IEEE Annals of the History of Computing*, 33(4): 5-16.
8. DeCost, B.L., Hattrick-Simpers, J.R., Trautt, Z., Kusne, A.G., Campo, E. and Green, M.L. 2020. Scientific AI in materials science: A path to a sustainable and scalable paradigm. *Machine Learning: Science and Technology*, 1(3): 033001.
9. Vasudevan, R.K., Choudhary, K., Mehta, A., Smith, R., Kusne, G., Tavazza, F. et al. 2019. Materials science in the AI age: High-throughput library generation, machine learning and a pathway from correlations to the underpinning physics. *MRS Communications*, 9(3).
10. Goswami, L., Deka, M.K. and Roy, M. 2023. Artificial intelligence in material engineering: A review on applications of artificial intelligence in material engineering. *Advanced Engineering Materials*, 25(13): 2300104.
11. Hinton, G., Deng, L., Yu, D., Dahl, G.E., Mohamed, A., Jaitly, N. et al. 2012. Deep neural networks for acoustic modeling in speech recognition: The shared views of four research groups. *IEEE Signal Processing Magazine*, 29(6): 82-97.
12. Park, H., Zhu, R., Huerta, E., Chaudhuri, S., Tajkhorshid, E. and Cooper, D. 2023. End-to-end AI framework for interpretable prediction of molecular and crystal properties. *Machine Learning: Science and Technology*, 4: 025036.
13. Kalidindi, S.R., Brough, D.B., Li, S., Cecen, A., Blekh, A.L., Congo, F.Y.P. et al. 2016. Role of materials data science and informatics in accelerated materials innovation. *MRS Bulletin*, 41(8): 596-602.
14. Reyes, K.G. and Maruyama, B. 2019. The machine learning revolution in materials? *MRS Bulletin*, 44(7): 530-537.
15. Samine, S., Zemzami, M., Hmina, N., Lagache, M. and Belhouideg, S. 2022. Towards the use of artificial intelligence and machine learning in material scientist field. *In: 2022 8th International Conference on Optimization and Applications (ICOA)*. IEEE.
16. Schmidt, J., Marques, M.R.G, Botti, S., and Marques, M.A.L. 2019. Recent advances and applications of machine learning in solid-state materials science. *NPJ Computational Materials*, 5(1): 83.
17. Stoll, A. and Benner, P. 2021. Machine learning for material characterization with an application for predicting mechanical properties. *GAMM-Mitteilungen*, 44(1): e202100003.
18. Bock, F.E., Aydin, R.C., Cyron, C.J., Huber, N., Kalidindi, S.R. and Klusemann, B. 2019. A review of the application of machine learning and data mining approaches in continuum materials mechanics. *Frontiers in Materials*, 6: 110.
19. Ramprasad, R., Batra, R., Pilania, G., Mannodi-Kanakkithodi, A. and Kim, C. 2017. Machine learning in materials informatics: Recent applications and prospects. *NPJ Computational Materials*, 3(1): 54.

20. Ward, L., Liu, R., Krishna, A., Hegde, V.I., Agrawal, A., Choudhary, A. et al. 2017. Including crystal structure attributes in machine learning models of formation energies via Voronoi tessellations. *Physical Review B*, 96(2): 024104.
21. Agrawal, A., Meredig, B., Wolverton, C. and Choudhary, A. 2016. A formation energy predictor for crystalline materials using ensemble data mining. *In: 2016 IEEE 16th International Conference on Data Mining Workshops (ICDMW)*. IEEE.
22. Choudhary, K., Garrity, K.F. and Tavazza, F. 2020. Data-driven discovery of 3D and 2D thermoelectric materials. *Journal of Physics: Condensed Matter*, 32(47): 475501.
23. Kirklin, S., Saal, J.E., Meredig, B., Thompson, A., Doak, J.W., Aykol, M. et al. 2015. The open quantum materials database (OQMD): Assessing the accuracy of DFT formation energies. *NPJ Computational Materials*, 1(1): 1-15.
24. Shen, J., Griesemer, S.D., Gopakumar, A., Baldassarri, B., Saal, J.E., Aykol, M. et al. 2022. Reflections on one million compounds in the open quantum materials database (OQMD). *Journal of Physics: Materials*, 5(3): 031001.
25. Saal, J.E., Kirklin, S., Aykol, M., Meredig, B. and Wolverton, C. 2013. Materials design and discovery with high-throughput density functional theory: The open quantum materials database (OQMD). *JOM*, 65: 1501-1509.
26. de Almeida, A.F., Moreira, R. and Rodrigues, T. 2019. Synthetic organic chemistry driven by artificial intelligence. *Nature Reviews Chemistry*, 3(10): 589-604.
27. Hadian, M., Seyed Mohammad Ebrahimi, Mohammadzadeh, A. and Babaei, M. 2021 Application of artificial intelligence in modeling, control, and fault diagnosis. *Applications of Artificial Intelligence in Process Systems Engineering*, Elsevier. 255-323.
28. Ihom, A. and Offiong, A. 2015. Neural networks in materials science and engineering: A review of salient issues. *European Journal of Engineering and Technology*, 3(7): 40-54.
29. Segler, M.H., Preuss, M. and Waller, M.P. 2018. Planning chemical syntheses with deep neural networks and symbolic AI. *Nature*, 555(7698): 604-610.
30. Zhang, W.Y. 2010. Artificial neural networks in materials science application. *Applied Mechanics and Materials*, 20: 1211-1216.
31. Hkdh, B. 1999. Neural networks in materials science. *ISIJ International*, 39(10): 966-979.
32. Dobrzański, L., Trzaska, J. and Dobrzańska-Danikiewicz, A. 2014. *Use of Neural Networks and Artificial Intelligence Tools for Modeling, Characterization, and Forecasting in Material Engineering.* Elsevier.
33. Chauhan, R., Ghanshala, K.K. and Joshi, R. 2018. Convolutional neural network (CNN) for image detection and recognition. *In: 2018 First International Conference on Secure Cyber Computing and Communication (ICSCCC)*. IEEE.
34. Sorini, A., Pineda, E.J., Stuckner, J. and Gustafson, P.A. 2021. A convolutional neural network for multiscale modeling of composite materials. *In: AIAA Scitech 2021 Forum*.
35. Kalliatakis, G., Stamatiadis, G., Ehsan, S., Leonardis, A., Gall, J., Sticlaru, A. et al. 2017. Evaluating deep convolutional neural networks for material classification. arXiv preprint arXiv:1703.04101.
36. Scarselli, F., Gori, M., Tsoi, A.C., Hagenbuchner, M. and Monfardini, G. 2008. The graph neural network model. *IEEE Transactions on Neural Networks*, 20(1): 61-80.
37. Ronchetti, C., Puccini, M., Ferlito, S., Giusepponi, S., Palombi, F., Buonocore, F. et al. 2022. Machine learning techniques for data analysis in materials science. *In: 2022 AEIT International Annual Conference (AEIT)*. IEEE.

38. Omee, S.S., Louis, S.Y., Fu, N., Wei, L., Dey, S., Dong, R. et al. 2022. Scalable deeper graph neural networks for high-performance materials property prediction. *Patterns*. 3: 100491.
39. Allotey, J., Butler, K.T. and Thiyagalingam, J. 2021. Entropy-based active learning of graph neural network surrogate models for materials properties. *The Journal of Chemical Physics*, 155(17).
40. Harshvardhan, G.M., Gourisaria, M.K., Pandey, M. and Rautaray, S.S. 2020. A comprehensive survey and analysis of generative models in machine learning. *Computer Science Review*, 38: 100285.
41. Butler, K.T., Davies, D.W., Cartwright, H., Isayev, O. and Walsh, A. 2018. Machine learning for molecular and materials science. *Nature*, 559(7715): 547-555.
42. Balachandran, P.V., Kowalski, B., Sehirlioglu, A. and Lookman, T. 2018. Experimental search for high-temperature ferroelectric perovskites guided by two-step machine learning. *Nature Communications*, 9(1): 1668.
43. Guo, K., Yang, Z., Yu, C.H. and Buehler, M.J. 2021. Artificial intelligence and machine learning in design of mechanical materials. *Materials Horizons*, 8(4): 1153-1172.
44. Fu, Z., Liu, W., Huang, C. and Mei, T. 2022. A review of performance prediction based on machine learning in materials science. *Nanomaterials*, 12(17): 2957.
45. Duan, C., Nandy, A. and Kulik, H.J. 2022. Machine learning for the discovery, design, and engineering of materials. *Annual Review of Chemical and Biomolecular Engineering*, 13: 405-429.
46. Artrith, N. and Kolpak, A.M. 2014. Understanding the composition and activity of electrocatalytic nanoalloys in aqueous solvents: A combination of DFT and accurate neural network potentials. *Nano Letters*, 14(5): 2670-2676.
47. Wang, Q., Wu, J.F., Lu, Z., Ciucci, F., Pang, W.K. and Guo, X. 2019. A new lithium-ion conductor LiTaSiO5: Theoretical prediction, materials synthesis, and ionic conductivity. *Advanced Functional Materials*, 29(37): 1904232.
48. Kusaba, M., Liu, C., Koyama, Y., Terakura, K. and Yoshida, R. 2021. Recreation of the periodic table with an unsupervised machine learning algorithm. *Scientific Reports*, 11(1): 4780.
49. Zhou, Q., Tang, P., Liu, S., Pan, J., Yan, Q. and Zhang, S.C. 2018. Learning atoms for materials discovery. *Proceedings of the National Academy of Sciences*, 115(28): E6411-E6417.
50. Seko, A., Togo, A. and Tanaka, I. 2018. Descriptors for machine learning of materials data. *Nanoinformatics*, 2018: 3-23.
51. Pattnaik, P., Raghunathan, S., Kalluri, T., Bhimalapuram, P., Jawahar, C.V. and Priyakumar, U.D. 2020. Machine learning for accurate force calculations in molecular dynamics simulations. *The Journal of Physical Chemistry A*, 124(34): 6954-6967.
52. Botu, V. and Ramprasad, R. 2015. Learning scheme to predict atomic forces and accelerate materials simulations. *Physical Review B*, 92(9): 094306.
53. Rodney, D., Tanguy, A. and Vandembroucq, D. 2011. Modeling the mechanics of amorphous solids at different length scale and time scale. *Modelling and Simulation in Materials Science and Engineering*, 19(8): 083001.
54. Mahadevan, S. and Zhao, Y. 2002. Advanced computer simulation of polycrystalline microstructure. *Computer Methods in Applied Mechanics and Engineering*, 191(34): 3651-3667.
55. Behler, J., 2010. Neural network potential-energy surfaces for atomistic simulations. *Chemical Modelling: Applications and Theory*, 7: 1-41.

56. Dickel, D., Nitol, M. and Barrett, C. 2021. LAMMPS implementation of rapid artificial neural network derived interatomic potentials. *Computational Materials Science*, 196: 110481.
57. Artrith, N., Hiller, B. and Behler, J. 2013. Neural network potentials for metals and oxides – First applications to copper clusters at zinc oxide. *Physica Status Solidi (B)*, 250(6): 1191-1203.
58. Kailkhura, B., Gallagher, B., Kim, S., Hiszpanski, A. and Han, T.Y.J. 2019. Reliable and explainable machine-learning methods for accelerated material discovery. *NPJ Computational Materials*, 5(1): 108.
59. Janet, J.P., Chan, L. and Kulik, H.J. 2018. Accelerating chemical discovery with machine learning: Simulated evolution of spin crossover complexes with an artificial neural network. *The Journal of Physical Chemistry Letters*, 9(5): 1064-1071.
60. Panapitiya, G., G., Avendano-Franco, G., Ren, P., Wen, X., Li, Y. and Lewis, J.P. 2018. Machine-learning prediction of CO adsorption in thiolated, Ag-alloyed Au nanoclusters. *Journal of the American Chemical Society*, 140(50): 17508-17514.
61. Gao, C., Min, X., Fang, M., Tao, T., Zheng, X., Liu, Y. et al. 2022. Innovative materials science via machine learning. *Advanced Functional Materials*, 32(1): 2108044.
62. Pilania, G., Wang, C., Jiang, X., Rajasekaran, S. and Ramprasad, R. 2013. Accelerating materials property predictions using machine learning. *Scientific Reports*, 3(1): 2810.
63. Zhang, K., Zhang, Y. and Wang, S. 2013. Enhancing thermoelectric properties of organic composites through hierarchical nanostructures. *Scientific Reports*, 3(1): 3448.
64. Tang, W., Zhang, J., Ratnasingham, S., Liscio, F., Chen, K., Liu, T. et al. 2020. Substitutional doping of hybrid organic-inorganic perovskite crystals for thermoelectrics. *Journal of Materials Chemistry A*, 8(27): 13594-13599.
65. Sang, S., Xu, C., Fan, J., Miao, D., Side, C. and Wang, Z. 2023. Accurate prediction of microstructure of composites using machine learning. *Advanced Theory and Simulations*, 6(2): 2200674.
66. Li, Z., Xu, Q., Sun, Q., Hou, Z. and Yin, W.J. 2019. Thermodynamic stability landscape of halide double perovskites via high-throughput computing and machine learning. *Advanced Functional Materials*, 29(9): 1807280.
67. Lu, S., Zhou, Q., Ouyang, Y., Guo, Y., Li, Q. and Wang, J. 2018. Accelerated discovery of stable lead-free hybrid organic-inorganic perovskites via machine learning. *Nature Communications*, 9(1): 3405.
68. Lyu, R., Moore, C.E., Liu, T., Yu, Y. and Wu, Y. 2021. Predictive design model for low-dimensional organic-inorganic halide perovskites assisted by machine learning. *Journal of the American Chemical Society*, 143(32): 12766-12776.
69. Sun, M., Wu, T., Xue, Y., Dougherty, A.W., Huang, B. and Li, Y. 2019. Mapping of atomic catalyst on graphdiyne. *Nano Energy*, 62: 754-763.
70. Zhuo, Y., Mansouri Tehrani, A., Oliynyk, A.O., Duke, A.C. and Brgoch, J. 2018. Identifying an efficient, thermally robust inorganic phosphor host via machine learning. *Nature Communications*, 9(1): 4377.
71. Kim, S., Lee, J.M., Park, J., Lee, C., Park, S., Seo, J. et al. 2021. Inverse design of organic light-emitting diode structure based on deep neural networks. *Nanophotonics*, 10(18): 4533-4541.
72. Gómez-Bombarelli, R., Aguilera-Iparraguirre, J., Hirzel, T.D., Duvenaud, D., Maclaurin, D., Blood-Forsynthe, M.A. et al. 2016. Design of efficient molecular organic light-emitting diodes by a high-throughput virtual screening and experimental approach. *Nature Materials*, 15(10): 1120-1127.

73. Matloob, R. and Soutchanski, M. 2016. Exploring organic synthesis with state-of-the-art planning techniques. *In: Proc. SPARK Workshop.*
74. Coley, C.W., Thomas III, D.A., Lummiss, J.A.M., Jaworski, J.N., Breen, C.P., Schultz, V. et al. 2019. A robotic platform for flow synthesis of organic compounds informed by AI planning. *Science*, 365(6453): eaax1566.
75. Szymkuć, S., Gajewska, E.P., Klucznik, T., Molga, K., Dittwald, P., Startek, M. et al. 2016. Computer-assisted synthetic planning: The end of the beginning. *Angewandte Chemie International Edition*, 55(20): 5904-5937.
76. Huo, H., Rong, Z., Kononova, O., Sun, W., Botari, T., He, T. et al. 2019. Semi-supervised machine-learning classification of materials synthesis procedures. *NPJ Computational Materials*, 5(1): 62.
77. Raccuglia, P., Elbert, K.C., Adler, P.D.F., Falk, C., Wenny, M.B., Mollo, A. et al. 2016. Machine-learning-assisted materials discovery using failed experiments. *Nature*, 533(7601): 73-76.
78. Sha, W., Guo, Y., Yuan, Q., Tang, S., Zhang, X., Lu, S. et al. 2020. Artificial intelligence to power the future of materials science and engineering. *Advanced Intelligent Systems*, 2(4): 1900143.
79. Sha, W. and Edwards, K. 2007. The use of artificial neural networks in materials science based research. *Materials & Design*, 28(6): 1747-1752.
80. Ong, S.P. 2019. Accelerating materials science with high-throughput computations and machine learning. *Computational Materials Science*, 161: 143-150.
81. Junming, Z., Weidong, Y. and Yan, L. 2021. Application of artificial intelligence in composite materials. *Advances in Mechanics*, 51(4): 865-900.
82. Suwardi, A., Wang, F.K., Xue, K., Han, M.Y., Teo, P., Wang, P. et al. 2022. Machine learning-driven biomaterials evolution. *Advanced Materials*, 34(1): 2102703.
83. Abolhasani, M., Brown, K.A. and Editors, G. 2023. Role of AI in experimental materials science. *MRS Bulletin*, 48(2): 134-141.
84. Nti, I.K., Adekoya, A.F., Weyori, B.A. and Nyarko-Boateng, O. 2022. Applications of artificial intelligence in engineering and manufacturing: A systematic review. *Journal of Intelligent Manufacturing*, 33(6): 1581-1601.
85. Zhang, H., Hong, E., Chen, X. and Liu, Z. 2023. Machine learning enables process optimization of aerosol Jet 3D printing based on the droplet morphology. *ACS Applied Materials & Interfaces*, 15(11): 14532-14545.
86. Hoogenboom, R. and Schubert, U.S. 2005. High-throughput synthesis equipment applied to polymer research. *Review of Scientific Instruments*, 76(6).
87. Hungenberg, K.-D. and Wulkow, M. 2018. *Modeling and Simulation in Polymer Reaction Engineering: A Modular Approach.* John Wiley & Sons.
88. Coley, C.W., Barzilay, R., Jaakkola, T.S. and Green, W.H. 2017. Prediction of organic reaction outcomes using machine learning. *ACS Central Science*, 3(5): 434-443.
89. Gao, H., Struble, T.J., Coley, C.W., Wang, Y., Green, W.H. and Jensen, K.F. 2018. Using machine learning to predict suitable conditions for organic reactions. *ACS Central Science*, 4(11): 1465-1476.
90. Caldwell, N.H.M., Breton, B.C., Holburn, D.M. and Young, T.C.W. 2003. Particle analysis using neural networks and image processing in the SEM. *Microscopy and Microanalysis*, 9(S02): 738-739.
91. Huang, B., Li, Z. and Li, J. 2018. An artificial intelligence atomic force microscope enabled by machine learning. *Nanoscale*, 10(45): 21320-21326.
92. Okunev, A.G., Mashukov, M.Y., Sankova, N.N., Nartova, A.V. and Matveev, A.V. 2021. Artificial intelligence for imaging data analysis in materials science: Microscopy

and behind. *In: IOP Conference Series: Materials Science and Engineering*. IOP Publishing.

93. Terry, J., Lau, M.L., Sun, J., Xu, C., Hendricks, B., Kise, J. et al. 2021. Analysis of extended X-ray absorption fine structure (EXAFS) data using artificial intelligence techniques. *Applied Surface Science*, 547: 149059.
94. Lee, J.W., Park, W.B., Lee, J.H., Singh, S.P. and Sohn, K.S. 2020. A deep-learning technique for phase identification in multiphase inorganic compounds using synthetic XRD powder patterns. *Nature Communications*, 11(1): 86.
95. Suzuki, Y. 2022. Automated data analysis for powder X-ray diffraction using machine learning. *Synchrotron Radiation News*, 35(4): 9-15.
96. Wang, H., Xie, Y., Li, D., Deng, H., Zhao, Y., Xin, M. et al. 2020. Rapid identification of X-ray diffraction patterns based on very limited data by interpretable convolutional neural networks. *Journal of Chemical Information and Modeling*, 60(4): 2004-2011.
97. Wang, X., Li, J., Ha, H.D., Dahl, J.C., Ondry, J.C., Moreno-Hernandez, I. et al. 2021. AutoDetect-mNP: An unsupervised machine learning algorithm for automated analysis of transmission electron microscope images of metal nanoparticles. *Jacs Au*, 1(3): 316-327.
98. Groschner, C.K., Choi, C. and Scott, M.C. 2020. Methodologies for successful segmentation of HRTEM images via neural network. arXiv e-prints, arXiv: 2001.05022.
99. Zhu, X., Mao, Y., Liu, J., Chen, Y., Chen, C., Li, Y. et al. 2023. Deep learning-assisted analysis of HRTEM images of crystalline nanoparticles. *Nanoscale*, 15: 14496-14504.
100. Groschner, C.K., Choi, C. and Scott, M.C. 2021. Machine learning pipeline for segmentation and defect identification from high-resolution transmission electron microscopy data. *Microscopy and Microanalysis*, 27(3): 549-556.

CHAPTER

2

Data Driven Artificial Intelligence Based Approach for the Determination of Structural Stress Distribution in ASTM D3039 Tensile Specimens of Carbon-Epoxy and Kevlar-Epoxy Based Composite Materials

Akshansh Mishra[1], Vijaykumar S. Jatti[2], Ashwini Vijaykumar Jatti[3*]

[1] School of Industrial and Information Engineering, Politecnico Di Milano, Milan, Italy
[2] Symbiosis Institute of Technology, Symbiosis International (Deemed University), Pune, India
[3] D.Y. Patil Institute of Technology, Savitribai Phule Pune University, Pune, India

1. Introduction

Composite materials, also known as composites, are a class of materials that are made up of two or more distinct components, each with unique properties. These components are combined to create a material with improved overall properties. The two main types of composites are particle-reinforced composites and fibre-reinforced composites.

Particle-reinforced composites consist of small particles, such as ceramic or metal, that are dispersed throughout a matrix material, such as plastic. These particles reinforce the matrix material and improve its strength and stiffness. Fibre-reinforced composites, on the other hand, consist of fibres, such as glass, carbon, or aramid, that are embedded in a matrix material. These fibres provide high strength and stiffness, while the matrix material provides toughness [1–5].

*Corresponding author: koti.ashwini@gmail.com

The history of composite materials dates back to ancient times when early civilizations used natural materials such as reeds, bark, and animal hair to make baskets and other objects. In the early 20th century, researchers began to experiment with synthetic materials, such as plastic and fiberglass, to create composite materials. During World War II, the use of composite materials in aircraft and other military equipment increased significantly. In the years that followed, the use of composite materials continued to grow, and by the 1960s, they were being used in a wide range of applications, including aerospace, automotive, construction, and sports equipment [6–8]. Today, composite materials are used in everything from airplanes and cars to bridges and buildings.

In recent years, the use of composite materials has continued to increase, driven by advances in technology and a growing need for lightweight, high-performance materials. The development of new composite materials, such as carbon-fibre composites, has made it possible to create even stronger and more lightweight materials.

One of the most significant advantages of composite materials is their high strength-to-weight ratio. The fibres or particles used in composite materials provide high strength and stiffness, while the matrix material provides toughness. This allows composite materials to be stronger than traditional materials of the same weight.

Another advantage of composite materials is their ability to resist corrosion. The fibres or particles used in composite materials are often resistant to corrosion, and the matrix material can also be formulated to provide additional protection against corrosion. This makes composite materials well-suited for use in harsh environments.

Composite materials also offer improved thermal stability, which can be beneficial in high-temperature applications. The fibres or particles used in composite materials can provide insulation against heat, and the matrix material can also be formulated to provide additional thermal protection. Composite materials also have excellent fatigue resistance, meaning that they can withstand repeated loading and unloading cycles without failure. This makes them ideal for use in applications such as aircraft and automotive parts, where parts are subjected to repeated loads during their lifetime. Composite materials can also be designed to have specific properties, such as electrical conductivity or thermal conductivity, by selecting the appropriate fibres or particles and matrix material. This allows engineers to tailor the properties of the composite material to suit specific applications.

Machine learning, a subset of artificial intelligence, is a powerful tool that has been applied to various fields, including composite materials. In recent years, there has been a growing interest in using machine learning techniques to improve the design, manufacturing, and characterization of composite materials.

One of the key applications of machine learning in composite materials is in the prediction of material properties. Researchers have used various techniques such as artificial neural networks, support vector machines, and Gaussian process regression to predict the mechanical properties of composite materials from their microstructures.

In order to make precise predictions about the macroscopic stiffness and yield strength of a unidirectional composite loaded in the transverse plane, Pathan et al. [9] applied data analytics and supervised machine learning. Without using physical calculations, predictions were made using a picture of the material's microstructure and information on the constitutive models of the fibres and matrix. The computational framework is built around analyzing the 2-point correlation function of the pictures of 1800 microstructures, principal component analysis, and dimensionality reduction. 1800 matching statistical volume elements (SVEs) depicting cylinder-shaped fibres in a continuous matrix loaded in the transverse plane were used in FE simulations.

In order to highlight the broad spectrum potential of ML in applications like prediction, optimization, feature identification, uncertainty quantification, reliability, and sensitivity analysis, along with the framework of various ML algorithms concerning polymer composites, Sharma et al. [10] condensed the findings of the extensive body of pertinent literature.

In order to calculate the relative crystallinity of biodegradable PLLA/PGA (polyglycolide) composites, Wang et al. [11] applied machine learning techniques. The relative crystallinity of PLLA/PGA polymer composites was calculated as a function of crystallization time, temperature, and PGA concentration using six different artificial intelligence classes.

The wear and friction of dry and lubricated aluminum-graphite composites, as well as the change between lubrication regimes, were studied by Hasan et al. [12]. In order to identify clusters in the friction and wear data, dimensionality reduction for the 14 material and tribological variables was carried out using Principal Component Analysis.

In order to obtain the precision of HOZT while having the low processing costs of FSDT, Mukhopadhyay et al. [13] proposed to use Gaussian process-based machine learning to build a computational bridge between the two. This article will use the term "modified FSDT" (mFSDT) to refer to the machine learning-augmented FSDT algorithm, on which detailed deterministic results and Monte Carlo simulation-assisted probabilistic results are presented for the free vibration analysis of shear deformation-sensitive structures like laminated composite and sandwich plates taking various configurations into account. The suggested method of connecting several laminate theories is generic in nature and can be applied further in a variety of additional static and dynamic assessments using composite plates and shells for precise but effective results.

Multi-objective optimizations were carried out by Wagner et al. [14] for the laminate stacking sequence of a composite cylinder under axial compression. Three different ply angles are used in the optimization along with several forms of geometric imperfections. Machine learning techniques based on decision trees were used to create general design suggestions that result in the largest buckling load and the least amount of flaw sensitivity.

A machine learning method for designing active composite structures that can produce desired shape-changing reactions was given by Hamel et al. [15]. They employed an evolutionary algorithm in conjunction with the finite element approach. The constructions of equally sized voxel units formed of either a passive or active material that optimize the distribution of these two material phases were put together in order to obtain a goal shape. To demonstrate the agreement between the desired shape and the best machine learning solution achieved, the optimization method was tested against a number of illustrative cases in active composite design.

2. Experimental Procedure

In the present work, two types of composite materials i.e., Carbon-Epoxy based, and Kevlar-epoxy based are used for designing the ASTM D3039 tensile specimens. The simulation work is carried out using ABAQUS CAE 2022 software. The design of ASTM D3039 is shown in Fig. 1.

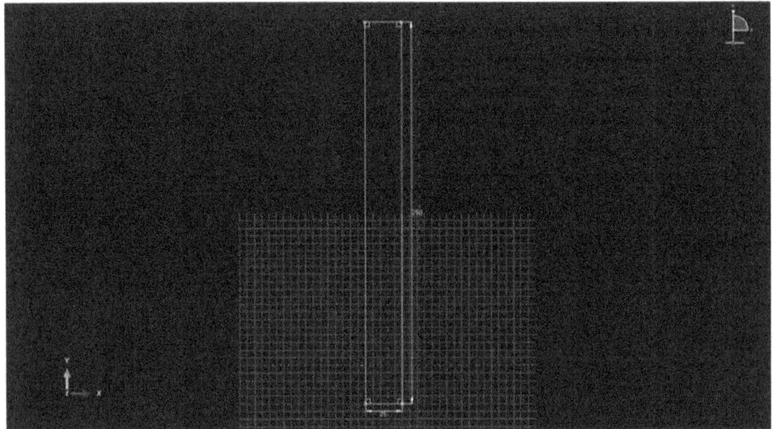

Figure 1: Dimensions of ASTM D3039 tensile specimen where length is 250 mm, width is 25 mm

Figure 2 shows the planar shell structure of the ASTM D3039 specimen. In the present work, the composite laminate consists of three-ply members of each 1 mm thickness. Tables 1 and 2 show the mechanical properties of the carbon-epoxy and Kevlar epoxy-based composite laminates. Table 3 shows the basic design of the simulation test carried out in the present work.

Table 1: Mechanical property of the carbon-epoxy laminate

E1	E2	Nu12	G12	G13	G23
138000	9500	0.28	5200	5200	1450

Figure 2: Planar shell structure of the ASTM D3039 specimen

Table 2: Mechanical property of the Kevlar-epoxy laminate

E1	E2	Nu12	G12	G13	G23
80000	5500	0.34	2200	2200	1800

The data from Table 3 is converted into a CSV file where Composite type, Ply 1 orientation, Ply 2 orientation, Ply 3 orientation, and Tensile Load (N) are input parameters while the von Mises stress (MPa) is an output parameter. The framework of the implemented machine learning framework used in the present work is shown in Fig. 3.

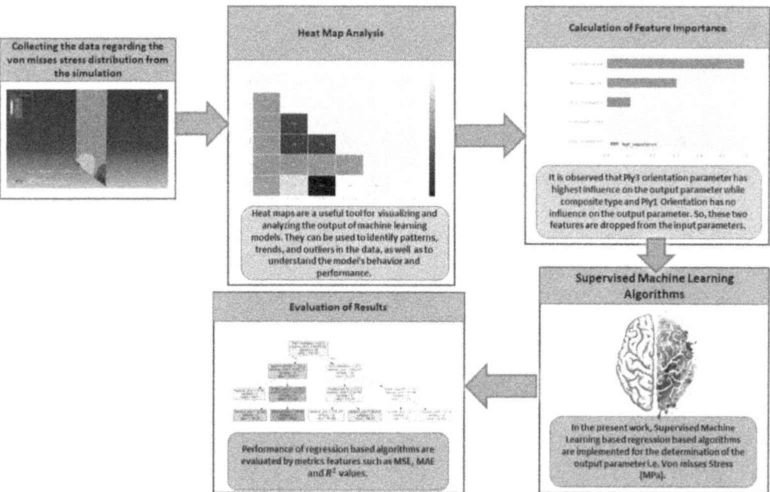

Figure 3: Machine learning framework to predict the von Mises stresses in the ASTM D3039 specimen

Table 3: Design of simulation test

Sample no.	Composite type	Ply1 orientation	Ply2 orientation	Ply3 orientation	Tensile load (N)	von Mises stress (MPa)
1	Carbon Epoxy	0	90	45	100	5.835
2	Carbon Epoxy	0	90	45	200	19.17
3	Carbon Epoxy	0	90	45	300	28.75
4	Carbon Epoxy	90	45	0	100	52.87
5	Carbon Epoxy	90	45	0	200	105.7
6	Carbon Epoxy	90	45	0	300	158.6
7	Carbon Epoxy	45	0	90	100	126.4
8	Carbon Epoxy	45	0	90	200	252.8
9	Carbon Epoxy	45	0	90	300	379.1
10	Kevlar Epoxy	0	90	45	100	8.747
11	Kevlar Epoxy	0	90	45	200	17.49
12	Kevlar Epoxy	0	90	45	300	26.24
13	Kevlar Epoxy	90	45	0	100	59.49
14	Kevlar Epoxy	90	45	0	200	119
15	Kevlar Epoxy	90	45	0	300	178.5
16	Kevlar Epoxy	45	0	90	100	133.6
17	Kevlar Epoxy	45	0	90	200	267.3
18	Kevlar Epoxy	45	0	90	300	400.9

Supervised machine learning is a method of training a model using labeled data, where the desired output is already known. Regression is a type of supervised machine learning algorithm that is used to predict a continuous value, such as a price or a quantity. The process of training a regression-based model begins with collecting a dataset of labeled data. The dataset should include the input variables (also known as features or predictors) and the output variable (also known as the target or response variable). Once the data is collected, it is split into a training set and a testing set. The training set is used to train the model, while the testing set is used to evaluate the performance of the model.

The next step is to choose a regression algorithm. There are several types of regression algorithms such as linear regression, polynomial regression, and logistic regression. Each algorithm has its own set of strengths and weaknesses, and the choice of algorithm will depend on the specific problem and dataset. After selecting the algorithm, the model is trained using the training set. During training, the model learns the relationship between the input variables and the output variable. The model is then tested on the testing set to evaluate its performance. If the performance is not satisfactory, the model can be fine-tuned by adjusting the parameters or by using a different algorithm.

Once the model is fine-tuned, it can be used to make predictions on new data. The model takes in new input data and uses the learned relationship to predict the output. The model's predictions can then be compared to the actual output to evaluate its performance.

3. Results and Discussion

Carbon epoxy composites are a type of advanced composite material that is widely used in aerospace, automotive, and other industries due to their high strength-to-weight ratio and excellent stiffness. These composites are made up of carbon fibres embedded in an epoxy matrix. The structural stress distribution in carbon epoxy composites is a complex phenomenon that is influenced by various factors such as the orientation and distribution of the carbon fibres, the properties of the epoxy matrix, and the loading conditions. One of the key factors that affect the structural stress distribution in carbon epoxy composites is the orientation of the carbon fibres. When the fibres are oriented in the direction of the applied load, the composite will exhibit high strength and stiffness in that direction. However, when the fibres are oriented in other directions, the composite will exhibit lower strength and stiffness in those directions.

Another important factor that affects the structural stress distribution in carbon epoxy composites is the distribution of carbon fibres. A more uniform distribution of fibres will result in a more homogeneous stress distribution throughout the composite. However, if the fibres are not evenly distributed, the stress will be concentrated in certain areas, leading to potential failure in those regions. The properties of the epoxy matrix also play a role in the stress distribution. A stiffer matrix will transfer more of the applied load to the fibres, leading to higher stress in the fibres. However, a more compliant matrix will absorb more of the load, leading to lower stress in the fibres.

Finally, the loading conditions also affect the stress distribution. For example, a composite subjected to a tensile load will experience a higher stress in the fibres along the length of the sample, while a composite subjected to a compressive load will experience a higher stress in the fibres along the width of the sample.

Figures 4a-4c shows the von Mises stress distribution in individual plies. From the results it is observed that Ply 1 is being subjected to the von Mises stress of 5.835 MPa, Ply 2 is being subjected to the von Mises stress of 46.17 MPa while Ply 3 is being subjected to the von Mises stress of 11.19 MPa.

Figures 5a-5c and Figs. 6a-6c shows the principal stress distribution i.e., S11 and S22 in individual plies.

From the S11 stress distribution results, it is observed that Ply 1, Ply 2 and Ply 3 are subjected to the stress of 0.48 MPa, 46.47 MPa, and 11.7 MPa respectively. From the S22 stress distribution results, it is observed that Ply 1, Ply 2, and Ply 3 are subjected to the stress of 3.15 MPa, 0.77 MPa, and 2.86 MPa.

Figure 7 shows the maximum strain distribution in sample 1. A thickness vs strain plot in Abaqus simulation of composite materials shows the relationship between the thickness of the composite laminate and the strain experienced by the material. The plot can be used to understand how the thickness of the laminate affects the overall strain distribution and can help optimize the laminate design for specific loading conditions.

Figure 8 shows the strain variation as per the thickness of the tensile sample 1. A thickness vs stress plot in Abaqus simulation of composite materials shows the relationship between the thickness of the composite laminate and the stress experienced by the material. The plot can be used to understand how the thickness of the laminate affects the overall stress distribution and can help optimize the laminate design for specific loading conditions. Figure 9 shows the stress variation as per the thickness of the tensile sample 1.

When a composite laminate is subjected to a load, the individual plies experience different strains and stresses due to the variations in their thickness. The strain and stress variation curves for composite laminates can be divided into three regions: the centre region, the middle region, and the edge region. In the centre region, the strains and stresses are highest and decrease as we move towards the edges. This is due to the fact that the centre plies are subjected to the highest loads and therefore experience the highest strains and stresses. In the middle region, the strains and stresses are relatively constant and do not vary significantly with thickness. This is because the middle plies are not subjected to as high loads as the centre plies and therefore experience less variation in strains and stresses. In the edge region, the strains and stresses are lowest and increase as we move towards the centre. This is due to the fact that the edge plies are subjected to the lowest loads and therefore experience the lowest strains and stresses.

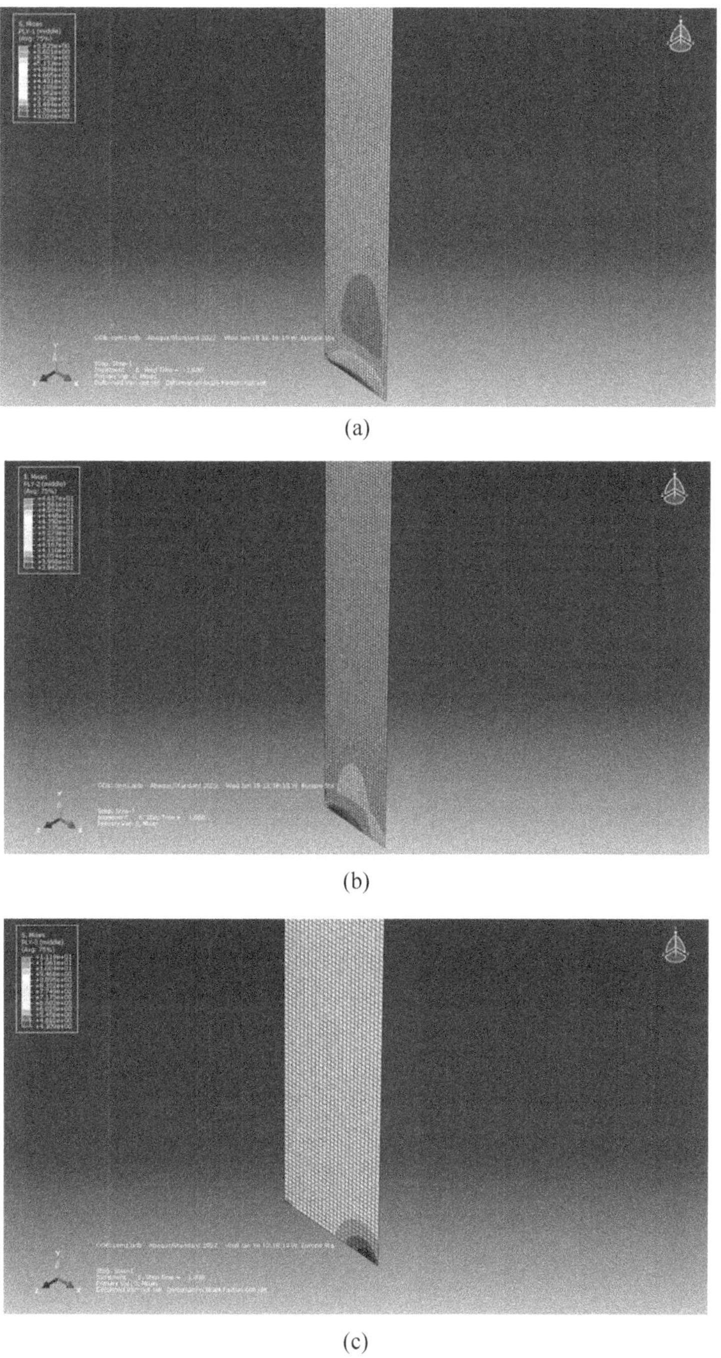

(a)

(b)

(c)

Figure 4: von Mises stress distribution in (a) Ply 1, (b) Ply 2 and (c) Ply 3 of sample 1

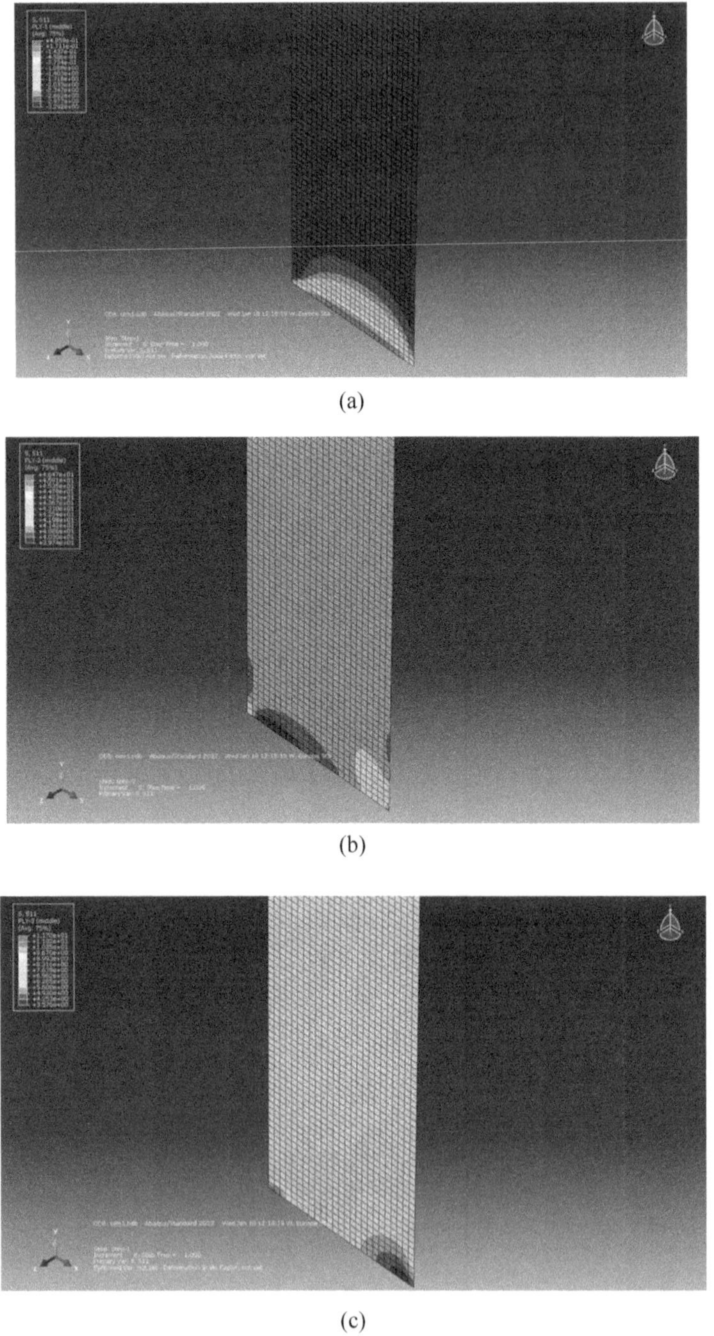

(a)

(b)

(c)

Figure 5: S11 stress distribution in (a) Ply 1, (b) Ply 2 and (c) Ply 3 of sample 1

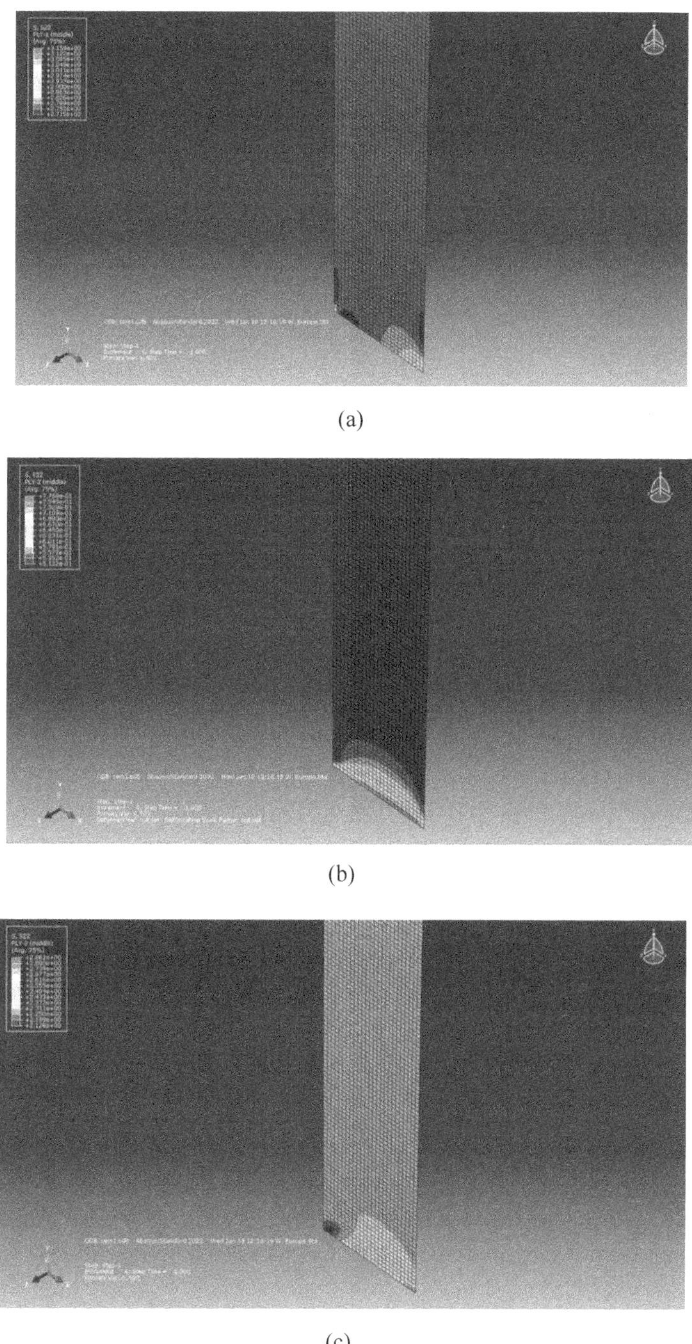

(a)

(b)

(c)

Figure 6: S22 stress distribution in (a) Ply 1, (b) Ply 2 and (c) Ply 3

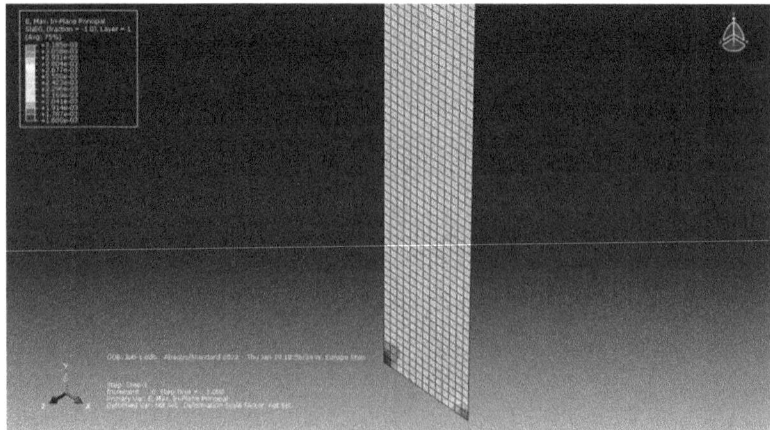

Figure 7: Maximum strain distribution in sample 1

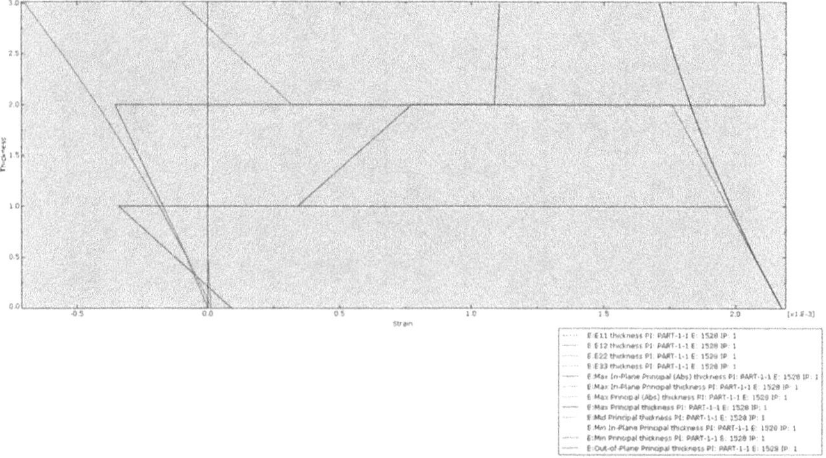

Figure 8: Variation of strains with thickness of laminate for sample 1

Figure 10a-10c shows the ply stack sequence plot for the given samples.

One of the main features of Kevlar-epoxy based composites is their layered structure, where the Kevlar fibres are embedded in an epoxy matrix. The structural distribution of Kevlar-epoxy based composites can be divided into three main regions: the fibre region, the matrix region, and the interface region. In the fibre region, the Kevlar fibres are the primary load-bearing component and are responsible for providing the composite with its high strength and toughness properties. The fibres are arranged in a specific orientation in order to achieve the desired mechanical properties. In the matrix region, the epoxy matrix surrounds and binds the fibres together, providing the composite with excellent adhesion and chemical resistance properties. The matrix also helps to distribute the loads

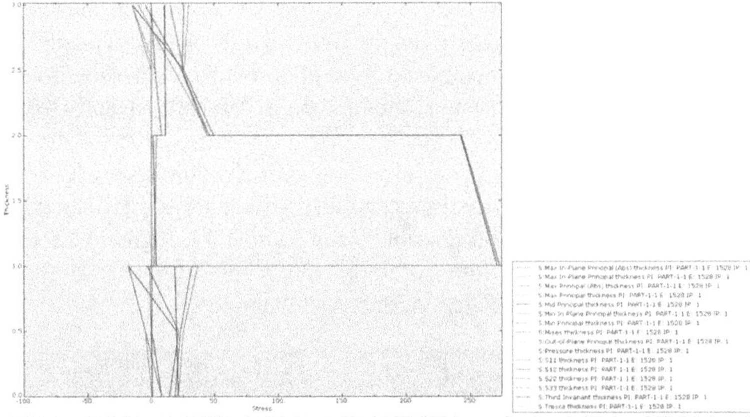

Figure 9: Variation of stress with thickness of laminate for sample 1

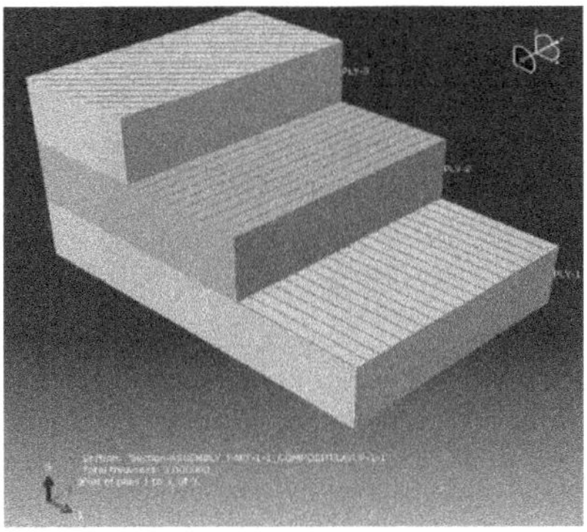

(a)

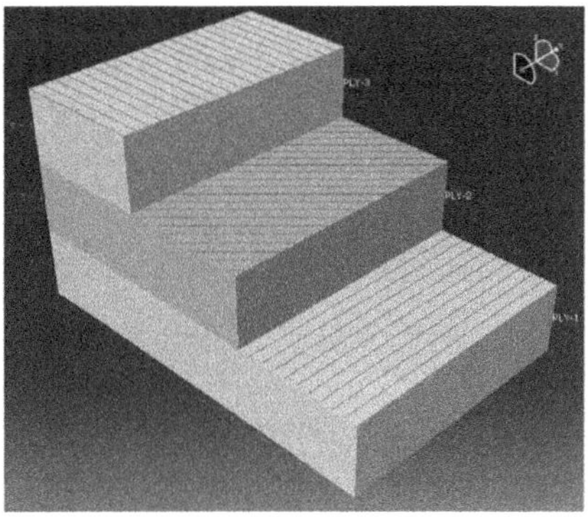

(b)

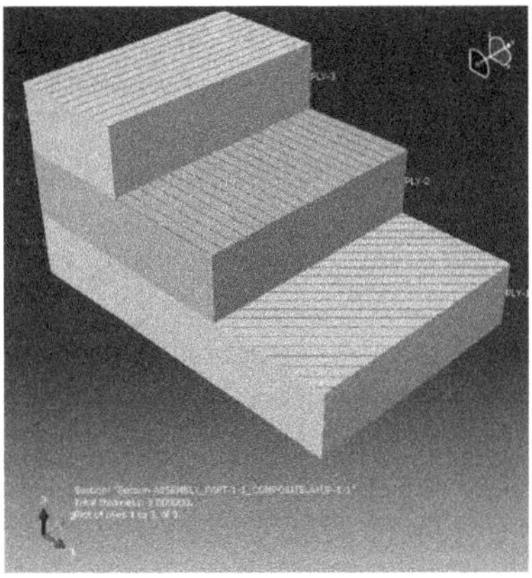

(c)

Figure 10: Ply sequence plot of (a) 0/90/45 samples, (b) 90/45/0 samples and (c) 45/0/90 samples

applied to the composite and protect the fibres from environmental damage. In the interface region, the interface between the fibres and the matrix is crucial for the mechanical properties of the composite. A good fibre-matrix adhesion is essential for the transfer of loads between the fibres and the matrix and also helps to prevent delamination and cracking.

For sample 13 the overall von Mises stress distribution is shown in Fig. 11. Figures 12a-12c shows the von Mises stress distribution in individual plies. Figure 13 shows the maximum strain distribution in sample 13. Figure 14 shows the strain variation as per the thickness of the tensile sample 13. Figure 15 shows the stress variation as per the thickness of the tensile sample 13.

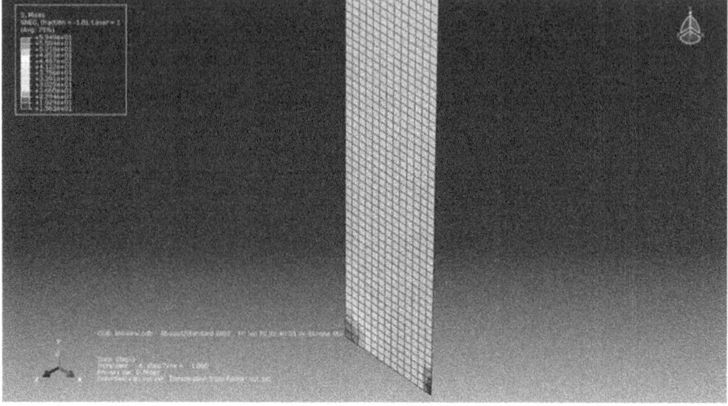

Figure 11: von Mises stress distribution in sample 13

Carbon-epoxy composites consist of carbon fibres embedded in an epoxy matrix. They are known for their high strength-to-weight ratio and excellent stiffness properties. The thickness vs stress curve for carbon-epoxy composites is generally linear, with the stress increasing proportionally with the thickness. However, as the thickness increases, the strength of the composite tends to plateau, indicating that there is a limit to the strength that can be achieved with a thicker composite.

Kevlar-epoxy composites, on the other hand, consist of Kevlar fibres embedded in an epoxy matrix. They are known for their high toughness and impact resistance properties. The thickness vs stress curve for Kevlar-epoxy composites is generally non-linear, with the stress increasing at a slower rate as the thickness increases. This is due to the fact that Kevlar fibres are highly flexible and can absorb large amounts of energy before breaking.

In terms of strength-to-weight ratio, carbon-epoxy composites tend to have a higher ratio than Kevlar-epoxy composites, due to their higher stiffness and strength properties. However, Kevlar-epoxy composites tend to have a higher toughness and impact resistance than carbon-epoxy composites, due to their non-linear thickness vs stress curve.

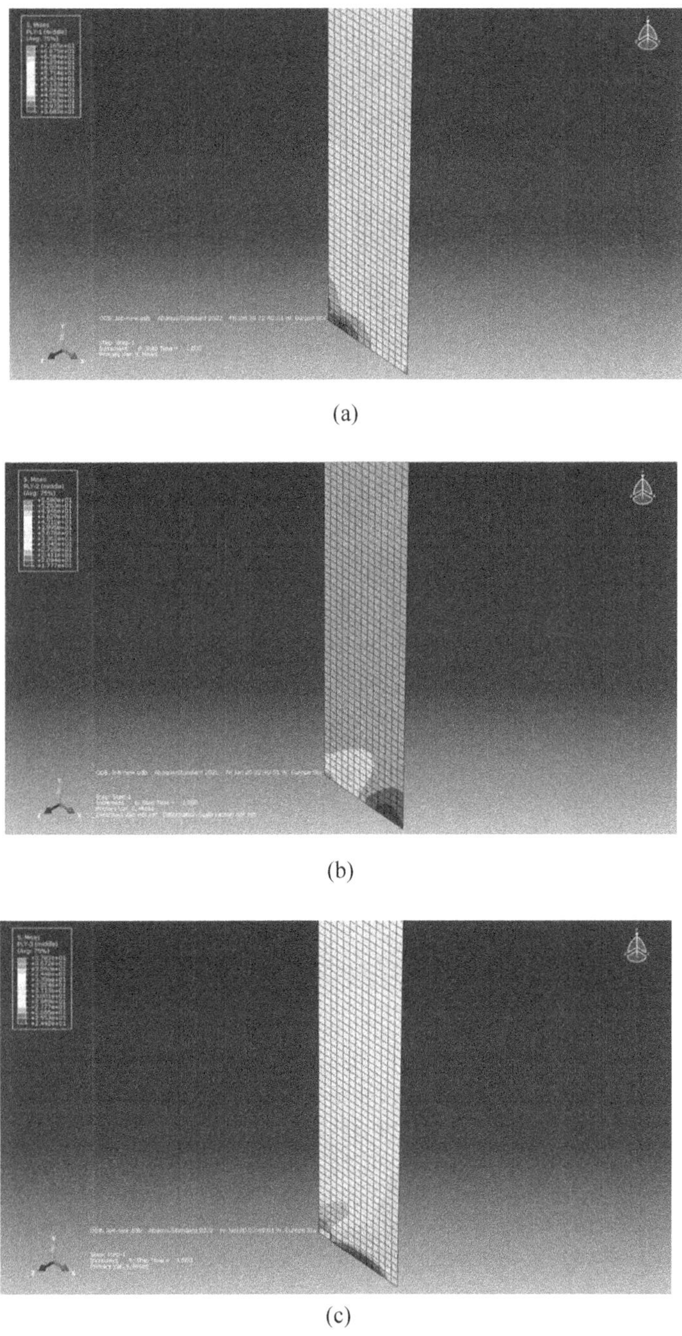

(a)

(b)

(c)

Figure 12: von Mises distribution in (a) Ply 1, (b) Ply 2 and (c) Ply 3 of sample 13

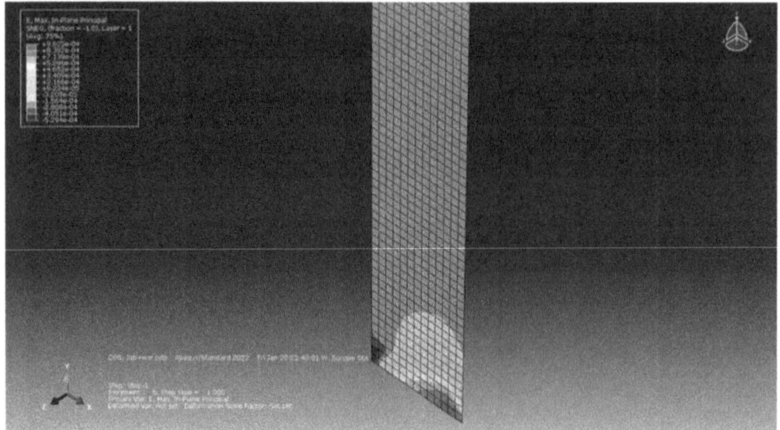

Figure 13: Maximum strain distribution in sample 13

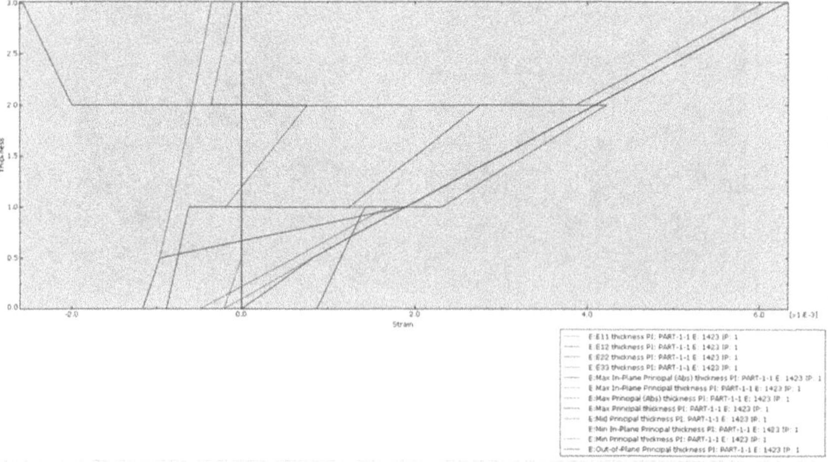

Figure 14: Strain plot obtained for sample 13

In conclusion, the comparison of carbon-epoxy composites and Kevlar-epoxy composites on the basis of thickness vs stress curves reveals that both types of composites have their own unique properties and characteristics. Carbon-epoxy composites tend to have a higher strength-to-weight ratio than Kevlar-epoxy composites, while Kevlar-epoxy composites tend to have a higher toughness and impact resistance. The choice of composite material will depend on the specific requirements of the application.

One of the main advantages of using composite materials in structural applications is their high strength-to-weight ratio, which allows for the construction of lightweight and strong structures. The von Mises stress is a key

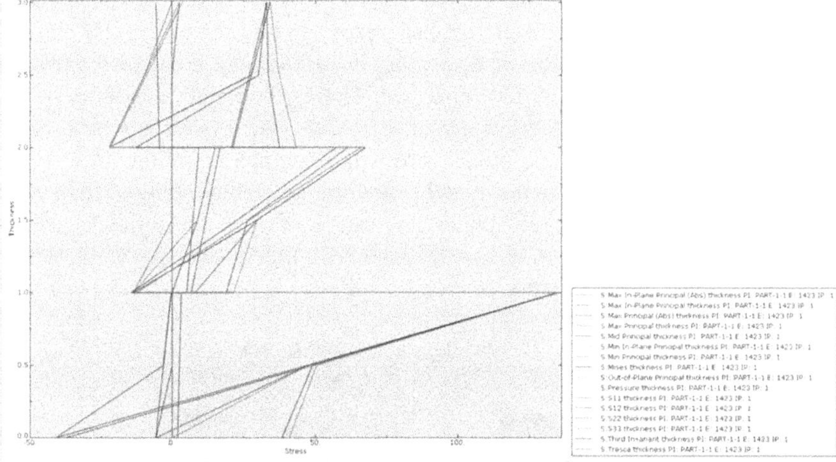

Figure 15: Stress plot obtained for sample 13

factor in determining the strength and performance of composite materials, as it is an indicator of the overall stress state of the material. High von Mises stress values indicate that the material is approaching its failure point, while low von Mises stress values indicate that the material has sufficient strength to withstand the applied loads. Another important aspect of composite materials is their ability to resist failure due to impact loads. The von Mises stress is also an important factor in determining the impact resistance of composite materials, as it provides a measure of the equivalent stress that the material is experiencing. High von Mises stress values indicate that the material is more likely to fail due to impact loads, while low von Mises stress values indicate that the material is more likely to resist impact loads. In addition to strength and impact resistance, von Mises stress also plays a crucial role in determining the fatigue life of composite materials. Fatigue failure occurs when a material is subjected to repeated loads, and the von Mises stress is an indicator of the stress state of the material during fatigue loading. High von Mises stress values indicate that the material is more likely to fail due to fatigue, while low von Mises stress values indicate that the material is more likely to have a longer fatigue life.

Heat maps are a powerful tool in machine learning for visualizing and understanding data. They can be used to display a wide range of information, including the distribution of data points, the correlation between variables, and the relationship between different features in a dataset. One of the main advantages of heat maps is their ability to display large amounts of data in a clear and concise manner. By using different colours to represent different values or ranges of values, heat maps can effectively convey complex information in a way that is easy to understand. This makes them particularly useful for exploring and analyzing large datasets. In addition to providing a clear visual representation of data, heat maps can also be used to identify patterns and trends that might not be immediately apparent from raw data. For example, heat maps can be used to identify clusters of data points that are similar or dissimilar, or to highlight areas of the dataset where certain variables are highly correlated. Another important use of heat maps in machine learning is in the evaluation of models. Heat maps can be used to visualize the performance of a model across different subsets of the data, which can help identify areas where the model is performing well or poorly. This can be very useful in understanding the underlying causes of model performance and in identifying areas where the model can be improved. Figure 16 shows the heat map plot obtained in the present work.

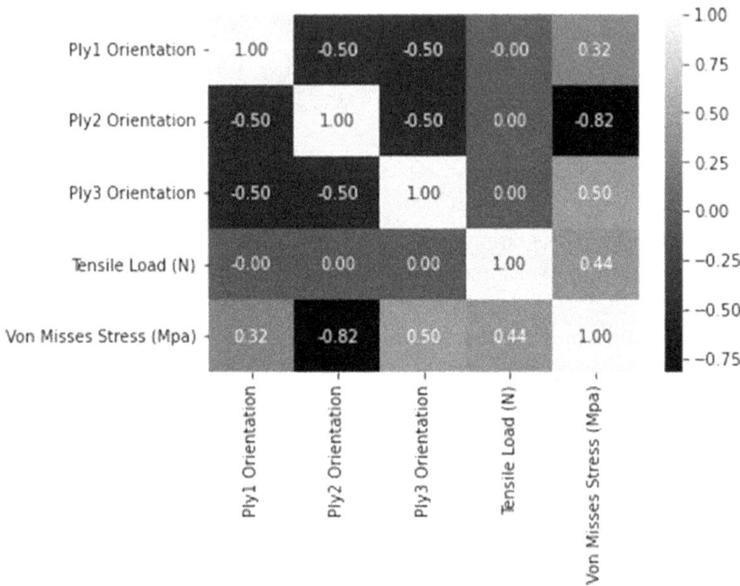

Figure 16: Heat map plot obtained in the present work

Feature importance plots are a common tool used in machine learning to understand which features of a dataset are most important in making predictions or classifications. These plots can be generated using various methods, such as decision trees, random forests, and gradient boosting, and they provide a way to identify which features are most important in a model's decision-making process. One of the main advantages of feature importance plots is that they can help identify which features are most relevant to the outcome of a model. This can be especially useful in datasets with many features, where it can be difficult to determine which features are most important for making predictions. By identifying the most important features, data scientists can focus their efforts on understanding and improving these features, rather than wasting time on less important features.

Another important use of feature importance plots is in feature selection. When building a machine learning model, it is often necessary to select a subset of features from a dataset to use as inputs. Feature importance plots can help identify which features are most important for a model's performance and can be used to guide the selection of features for a model. This can lead to models that are more accurate and efficient, as they are based on the most relevant features of the dataset.

In addition, feature importance plots can also be used to identify potential issues with a model, such as overfitting or bias. For example, if a feature importance plot shows that a single feature is responsible for a large proportion of the model's predictions, it may indicate that the model is overfitting to that feature. Similarly, if a feature importance plot shows that a certain feature is disproportionately important for one class of examples, it may indicate that the model is biased towards that class. Figure 17 shows the feature importance plot obtained in the present work.

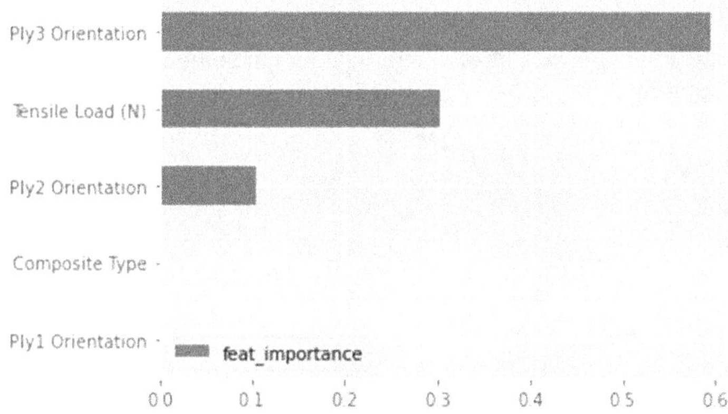

Figure 17: Feature importance plot obtained in the present work.

Table 4 shows the results of metrics features i.e., Mean Square Error, Mean Absolute Error and R^2 values for measuring the performance of the implemented machine learning algorithm. Figure 18 shows the obtained decision tree plot in the present work.

Table 4: Performance measurement of the implemented machine learning algorithms

Algorithms	MSE	MAE	R^2 value
Linear Regression	2732.38	37.02	0.8742
Decision Trees	193.052	12.769	0.9911
XG Boost	143.85	10.921	0.9933
Extra Tree Regressor	132.561	7.718	0.9938

A decision tree is a widely used machine learning algorithm that is used for both classification and regression tasks. The algorithm creates a tree-like model of decisions and their possible consequences. The tree is created by recursively splitting the data based on the most informative feature, which is determined by an impurity measure such as entropy or Gini impurity. The mechanism of a decision tree plot is based on the concept of recursive partitioning. The tree is built by repeatedly splitting the data into smaller subsets based on the values of the input features. The process starts at the root node, which represents the entire dataset. At each node, the algorithm selects the feature that results in the highest reduction in impurity. Once a feature is selected, the data is split into subsets based on the values of that feature. This process is repeated for each subset until a stopping criterion is met. The final decision tree is composed of internal nodes, which represent the features used for splitting, and leaf nodes, which represent the final predictions. The internal nodes are labelled with the feature used for splitting, and the leaf nodes are labelled with the predicted class or value. Each path from the root to a leaf node represents a decision rule. The decision tree plot can be used to interpret the model and understand how the decisions are made. The feature's importance can be determined by the number of times a feature is used for splitting.

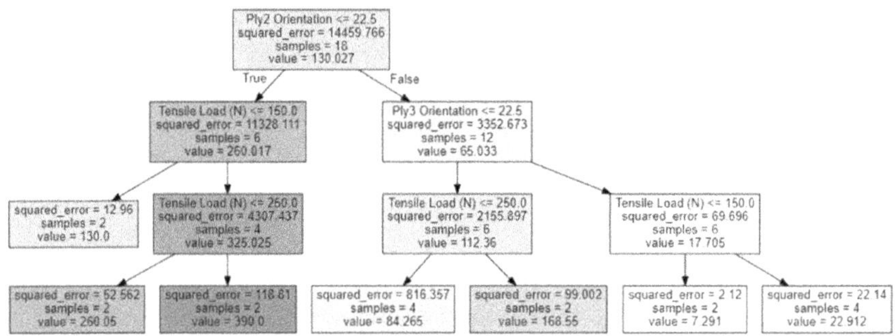

Figure 18: Decision tree plot

It is observed from Table 4 that the Extra Tree Regressor algorithm can predict the von Mises stress distribution with the highest accuracy.

Extra Tree Regressor is a powerful machine learning algorithm that is used for regression tasks. It is an ensemble method that builds multiple decision trees and combines their output to make a final prediction. The algorithm is known for its high accuracy and ability to handle large datasets. One of the key features of the Extra Tree Regressor is its randomness. Unlike traditional decision tree algorithms, Extra Trees randomly selects the feature to split on at each node. This randomness helps to reduce overfitting and improve the generalization of the model. Additionally, Extra Trees also uses random thresholds for each feature, which further increases its randomness and helps to improve its accuracy. Another feature that contributes to the high accuracy of the Extra Tree Regressor is its ability to handle missing data. The algorithm is able to handle missing values by randomly selecting a subset of the available features at each split. This allows the model to make accurate predictions even when there is missing data. Extra Trees also has a very fast training time, which is another advantage of the algorithm. The algorithm is able to handle large datasets quickly and efficiently, which makes it a good choice for large-scale regression tasks.

4. Conclusion

In conclusion, Artificial Intelligence (AI) based algorithms have been proven to be effective in the prediction of von Mises stresses in ASTM D3039 composite specimens. The results of this study demonstrate that AI-based methods such as XG Boost and Extra Tree Regressor can achieve high accuracy in predicting von Mises stresses in composite materials. The ability of AI-based algorithms to handle large datasets and complex relationships between input features and output variables makes them suitable for predicting von Mises stresses in composite specimens. The key conclusions of the present work are:

- The use of AI-based algorithms for the prediction of von Mises stresses in composite specimens is a promising approach. The results of this study demonstrate that AI-based methods can achieve high accuracy in predicting von Mises stresses in composite materials.
- The Extra Tree Regressor algorithm is a powerful machine learning technique that is known for its high accuracy and ability to handle large datasets. Its randomness and ability to handle missing data contribute to its high accuracy. Additionally, its fast training time makes it a good choice for large-scale regression tasks.

The selection of the appropriate algorithm, the appropriate parameters and the relevant features is crucial for achieving high-accuracy predictions. Future research should focus on the optimization of the AI-based algorithms and their parameters to improve the accuracy of the predictions.

References

1. Sun, G., Chen, D., Zhu, G. and Li, Q. 2022. Lightweight hybrid materials and structures for energy absorption: A state-of-the-art review and outlook. *Thin-Walled Structures*, 172: 108760.
2. Talreja, R. 2022. Damage and failure of composite materials. *In:* Advanced Theories for Deformation, Damage and Failure in Materials (pp. 235-280). Cham: Springer International Publishing.

3. Deshmukh, S.P., Shrivastava, R. and Thakar, C.M. 2022. Machining of composite materials through advance machining process. *Materials Today: Proceedings*, 52: 1078-1081.
4. Talreja, R. and Waas, A.M. 2022. Concepts and definitions related to mechanical behavior of fiber reinforced composite materials. *Composites Science and Technology*, 217: 109081.
5. Ambaye, T.G., Vaccari, M., Prasad, S., van Hullebusch, E.D. and Rtimi, S. 2022. Preparation and applications of chitosan and cellulose composite materials. *Journal of Environmental Management*, 301: 113850.
6. Singh, S., Uddin, M. and Prakash, C. 2022. Introduction, history, and origin of composite materials. *In:* Fabrication and Machining of Advanced Materials and Composites (pp. 1-18). CRC Press.
7. Yeswanth, I.V.S., Jha, K., Bhowmik, S., Kumar, R., Sharma, S. and Rushdan, A.I. 2022. Recent developments in RAM based MWCNT composite materials: A short review. *Functional Composites and Structures*, 4(2). DOI 10.1088/2631-6331/ac5730
8. Dave, H.K. 2023. Manufacturing and Characterization of Composites (Doctoral dissertation). Department of Mechanical Engineering, Sardar Vallabhbhai National Institute of Technology.
9. Pathan, M.V., Ponnusami, S.A., Pathan, J., Pitisongsawat, R., Erice, B., Petrinic, N. and Tagarielli, V.L. 2019. Predictions of the mechanical properties of unidirectional fibre composites by supervised machine learning. *Scientific Reports*, 9(1): 1-10.
10. Sharma, A., Mukhopadhyay, T., Rangappa, S.M., Siengchin, S. and Kushvaha, V. 2022. Advances in computational intelligence of polymer composite materials: Machine learning assisted modeling, analysis and design. *Archives of Computational Methods in Engineering*, 1-45.
11. Wang, J., Ayari, M.A., Khandakar, A., Chowdhury, M.E., Uz Zaman, S.A., Rahman, T. and Vaferi, B. 2022. Estimating the relative crystallinity of biodegradable polylactic acid and polyglycolide polymer composites by machine learning methodologies. *Polymers*, 14(3): 527.
12. Hasan, M.S., Kordijazi, A., Rohatgi, P.K. and Nosonovsky, M. 2022. Machine learning models of the transition from solid to liquid lubricated friction and wear in aluminum-graphite composites. *Tribology International*, 165: 107326.
13. Mukhopadhyay, T., Naskar, S. and Dey, S. 2023. On machine learning assisted data-driven bridging of FSDT and HOZT for high-fidelity uncertainty quantification of laminated composite and sandwich plates. *Composite Structures*, 304: 116276.
14. Wagner, H.N.R., Köke, H., Dähne, S., Niemann, S., Hühne, C. and Khakimova, R. 2019. Decision tree-based machine learning to optimize the laminate stacking of composite cylinders for maximum buckling load and minimum imperfection sensitivity. *Composite Structures*, 220: 45-63.
15. Hamel, C.M., Roach, D.J., Long, K.N., Demoly, F., Dunn, M.L. and Qi, H.J. 2019. Machine-learning based design of active composite structures for 4D printing. *Smart Materials and Structures*, 28(6): 065005.

CHAPTER

3

Image Segmentation for Evaluating the Microstructure Features obtained from Magnesium Composites Processed through Squeeze Casting

Anish Dasgupta[1*], Dhrubajyoti Sadhukhan[2]

[1] Consultant – Data Science/Engineer, Volvo Groups, Bangalore, India

[2] School of Industrial and Information Engineering, Politecnico Di Milano, Milan, Italy

1. Introduction

Modern composite materials known as "metal matrix composites" (MMCs) are made of a metallic matrix that has been reinforced with a secondary phase, such as ceramics or carbon. MMCs seek to combine the hardness, stiffness, and strength of the reinforcement material with the ductility and toughness of the metal matrix. The reinforcements often comprise particles, short fibres, continuous fibres, or whiskers made of materials like silicon carbide, boron carbide, alumina, or carbon nanotubes, while the metallic matrix is typically a lightweight metal like aluminium, magnesium, or titanium [1–6]. The use of the reinforcement material enhances the composite's physical and mechanical characteristics, such as its strength, stiffness, hardness, and high-temperature performance. To meet the requirements of the intended application, the metallic matrix, reinforcement material, shape, size, volume percentage, and orientation of the reinforcement phase are all carefully chosen. MMCs are produced utilizing methods that enable close mixing of the matrix and reinforcement, such as powder metallurgy, stir casting, squeeze casting, and molten metal infiltration. MMCs are appropriate for components in the aerospace, defense, automotive, and other demanding industries as a result of a class of light, strong, and stiff materials that can function

*Corresponding author: Anishdg27@gmail.com

at high temperatures [7–15]. However, difficulties in MMC production, joining, and machining persist, which may prevent their wide implementation. The goal of ongoing research is to increase the characteristics and manufacturability of these high-performance composites to enable wider application.

Machine learning, a subtype of artificial intelligence, allows computers to learn from experience and get better over time without having explicit programming [16–21]. Machine learning is being used to enhance workflows, evaluate data, and optimize designs in composites and additive manufacturing [22–28]. To increase composite strength, decrease flaws, and improve quality control, machine learning can assist manufacturers in optimizing manufacturing variables including curing time, temperature, and pressure. By identifying defects and irregularities, it provides automated feedback and real-time monitoring of production lines. Based on information from microstructural pictures, manufacturing conditions, and simulations, machine learning algorithms can also forecast the mechanical properties of composite materials. The creation of high-performance composites for demanding applications is made easier as a result. Machine learning enhances production quality in additive manufacturing and other industrial processes by managing process variables, modeling process-defect correlations, and evaluating printed products [29–35]. Utilizing generative design methodologies also shows potential for optimizing designs. Algorithms may be trained to anticipate part performance and find improvements using large datasets from simulations, previous builds, and in-situ monitoring. By enabling real-time process control, certification, and optimization, machine learning improves composites and additive manufacturing. It offers a data-driven strategy to hasten the creation of materials, improve designs, increase part consistency, and cut prices. For additive manufacturing and composites to reach their full potential in the future, machine learning and artificial intelligence must continue to progress.

Analyzing the microstructure of composite materials from microscope pictures can be done very well using the digital image processing technique known as image segmentation. In composite micrographs, segmentation algorithms may recognize and categorize various phases including pores, particles, fibres, and matrix regions. Pixels about each unique phase in the microstructure can be separated using common techniques such as thresholding, edge-based, region-based, and clustering algorithms. Important segmentation factors including gradient orientation, hue, and pixel intensity can distinguish reinforcing materials like silicon carbide or carbon fibres from the surrounding polymer, metal, or ceramic matrix. Quantitative data can be extracted after segmentation, including the distribution of fibre orientation, interfacial characteristics, defect counts, volume fractions of various phases, and nearest neighbor distances. The linkages between composite processing-structure-property relationships and quality control are clarified by the data. Correlations can be discovered by analyzing massive datasets of micrographs using cutting-edge segmentation algorithms for machine learning. In addition, segmentation makes it possible to create realistic virtual microstructure models for complex 3D simulations of composite behavior.

2. Materials and Methods

Stir casting and hot extrusion were used to create the metal matrix composites used in this investigation. Magnesium, zinc, and aluminum ingots made up the matrix, weighing 950–990 g magnesium per specimen. To melt and homogenize the alloy, the matrix ingots were warmed to 650°C and then heated to 850°C while being stirred. The carbon nanotubes (CNTs) and boron carbide (B4C) powder reinforcement phase were individually warmed to 300°C. The CNTs enhanced strength and stiffness while the B4C powder added hardness and wear resistance. To disperse the particles equally, the warmed reinforcement powders were added to the molten magnesium-zinc-aluminum matrix alloy and stirred for 10 minutes at various rpms. Once the mixture was uniform, it was conveyed with the aid of a runner system into a squeeze casting die that had been preheated to 300°C. A hydraulic press applied 40.2 tonnes of pressure to the die where it was compressed, held for 10 minutes, and then cooled. The metal matrix composite billet was then extruded to enhance the mechanical characteristics and particle distribution. The finished composite samples were created by machining and finishing the extruded rod. In comparison to the unreinforced alloy, the processing parameters were tuned to create composites with fewer casting flaws and better mechanical performance. The B4C and CNT particles were evenly distributed throughout the metal matrix thanks to stir casting and hot extrusion. With different matrix alloy compositions and reinforcements, three metal matrix composite specimens were created. In specimen 1, a carbon nanotube (CNT) matrix supplemented with 0.3% titanium carbide (TiC) and 1.5% carbon consisted of a magnesium AZ71 alloy matrix. The AZ71 alloy has 1% zinc and 7% aluminum. In specimen 2, the matrix was a magnesium alloy supplemented with 2% boron carbide (B4C) and 2% CNTs. The matrix contained 12% aluminum and 1% zinc. A magnesium alloy matrix made of 14% aluminum, 1% zinc, 2% B4C and 2% CNTs was also present in specimen 3. The TiC and B4C ceramic particle reinforcements improved hardness and wear resistance while the metallic matrices offered ductility. Strength and stiffness were improved by the CNTs. By using squeeze casting to create all of the composites, the molten magnesium alloy was able to penetrate the particle reinforcements. SRM, or specifically reinforced magnesium, is the name given to the composites. The mechanical properties and microstructure could be tailored for different applications requiring lightweight metal matrix composites with superior strength and stiffness compared to unreinforced magnesium alloys by varying the matrix alloy proportions of aluminium and zinc as well as the type, size, and quantity of reinforcements. Figures 1 and 2 show the specimens before and after finishing.

Figure 1: Specimen before finishing

Figure 2: Specimen after finishing

3. Results and Discussion

Before microscopic imaging, the appropriate surface preparation was carried out in order to clearly expose the microstructural characteristics and grain boundaries of the manufactured metal matrix composite specimens. The lathe left microscopic scratches and grooves on the objects' surfaces, despite the fact that they may have looked smooth to the human eye after being machined. Using successively finer grit emery papers, mechanical polishing was used to eliminate these flaws. 800, 1000, 1200, and 1500 grade emery papers were manually used to polish the surface in order to gradually remove machining marks and improve the smoothness of the surface finish. After achieving a surface devoid of deformation, chemical etching was carried out to target the grain boundaries with preference and improve their visibility under the microscope. 92 ml of distilled water, 6 ml of nitric acid with a 65% concentration, and 2 ml of hydrofluoric acid with a 40% concentration made up the etchant solution. While hydrofluoric acid helped show the reinforcement particles and their interfaces, nitric acid preferred to dissolve the matrix. The microstructure was examined using an optical or scanning electron microscope at different magnifications after the appropriate amount of etching time. Thus, damaged surface layers were removed and the contrast between the matrix, reinforcements, and boundaries was improved thanks to the combination of mechanical polishing and chemical etching. This allowed for a clear microstructural analysis of the metal matrix composites. Figures 3 and 4 shows the specimens before and after polishing.

The three manufactured metal matrix composite specimens differed significantly from one another according to the microstructural examination. The highly coarse-grained specimen 1 depicted in Fig. 5 had the potential for significant sliding at the grain boundaries. The matrix has an uneven distribution

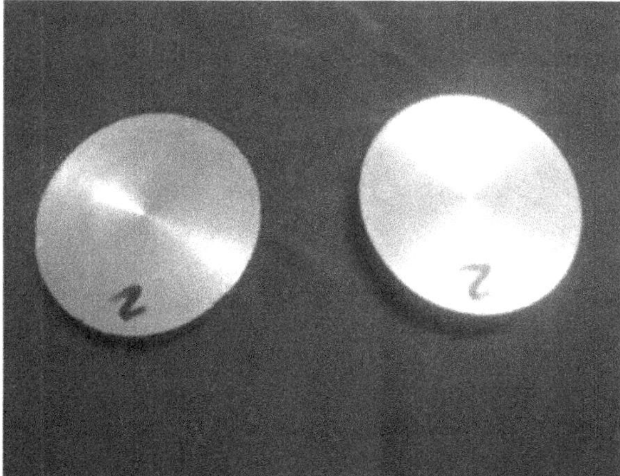

Figure 3: Unpolished and non-etched surface

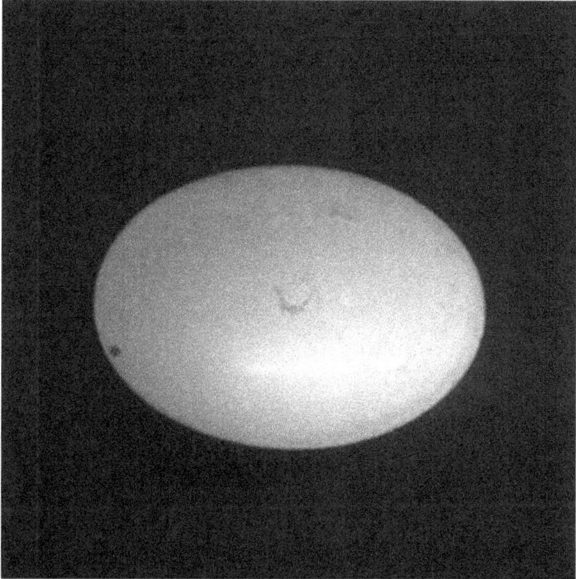

Figure 4: Specimen after emery polishing and etching

of the reinforcement particles. In contrast, the microstructure of specimen 2 in Fig. 6 showed a fine-grained refinement and homogeneous distribution of the reinforcement phases. There is less risk of grain boundary slippage with this microstructure. Figure 7 depicts specimen 3 with coarse granules that are coarser than specimens 1 and 2, but finer than specimen 3. Grain boundary slippage may be possible, but to a lesser extent than in specimen 1. In specimen 3, the reinforcement distribution was a little lopsided. The three specimens' different

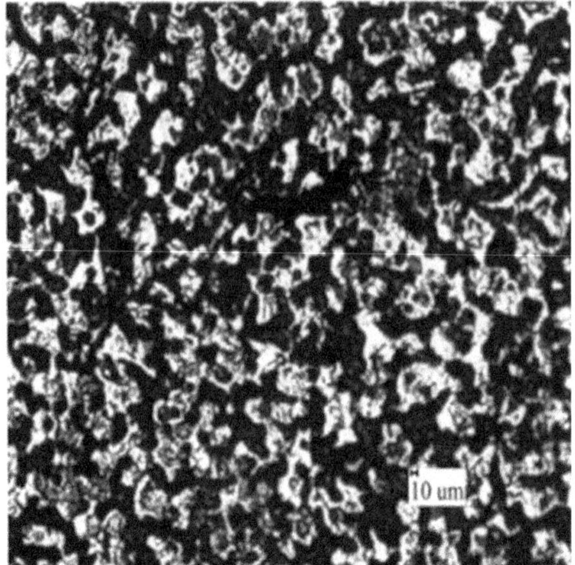

Figure 5: Magnification of 100X-SRM AL7Z1

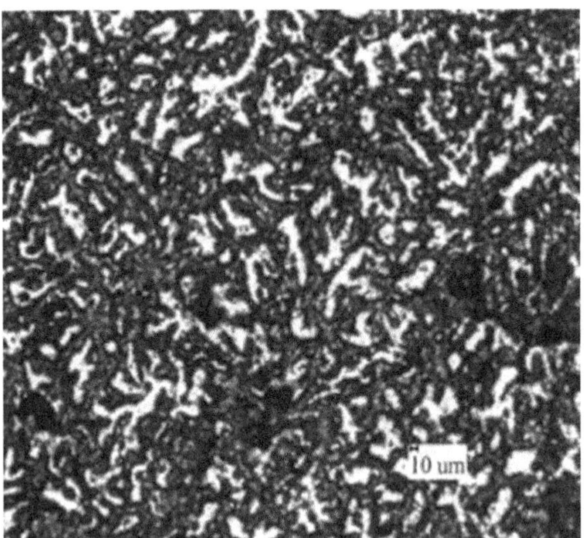

Figure 6: Magnification of 100X-SRM AL12Z1

matrix compositions and reinforcement types/fractions are to blame for the variances in grain size. The most homogenous microstructure, with finely scattered reinforcements, lowest porosity, and no intermetallic phases, was shown by specimen 2. According to the quantitative assessments, this ideal microstructure

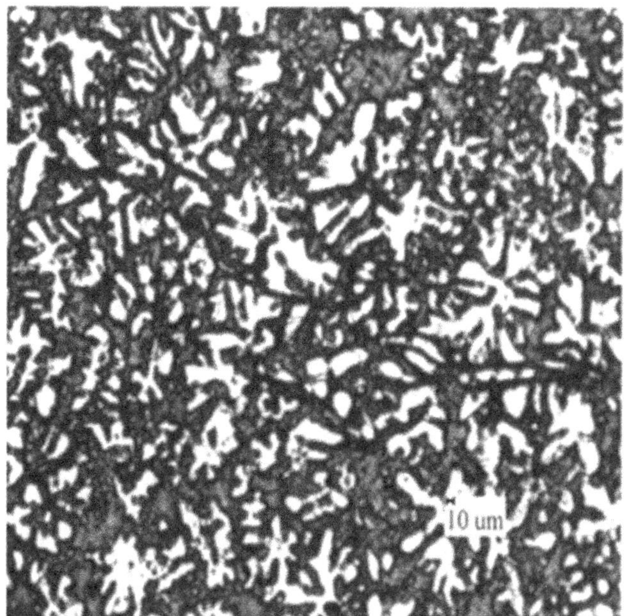

Figure 7: Magnification of 100X-SRM AL14Z1

gave specimen 2 superior hardness, strength, and ductility compared to the other two.

Active contours, also referred to as snakes, are flexible methods of image segmentation that adapt energy-saving curves to the edges of objects in an image. They are active because, based on the optimization of an energy function, the curves adapt iteratively to fit the required features. The energy function consists of internal forces that regulate curve smoothness and exterior forces that draw the curve to lines and edges in the image. For added sturdiness, other energy words can be used. In order to lock onto salient picture characteristics, active contours establish a curve near the object to be segmented and iteratively change its shape. Typically, partial differential equations and calculus of variations are used to shift the curve by minimizing the energy function. Shape priors and restrictions can be used in active contour models to increase accuracy. When object boundaries are diffuse or hazy, they function better than rigid edge detection, can segment complex shapes, and are helpful. Performance, however, is influenced by initial curve positioning and choosing the right energy terms. Multiple biological images, including those of cells, tissues, and anatomical structures, have been segmented using active contours. They are pertinent for composite microstructures with phase boundaries that are amorphous, overlapping, and fuzzy. In order to precisely extract shape information from such photos, active contours overcome the shortcomings of simpler segmentation techniques.

The grain boundaries and reinforcing stages in the micrographs of all three specimens could be clearly distinguished using the active contour segmentation technique. The coarse-grained microstructure of sample 1 in Fig. 8 resulted in a contour with more irregularities and complex boundaries that matched big grain shapes and unevenly distributed reinforcements. Figure 9 depicts the active contour for the fine-grained sample 2, in contrast, smoothly encircling the homogeneous and compact microconstituents.

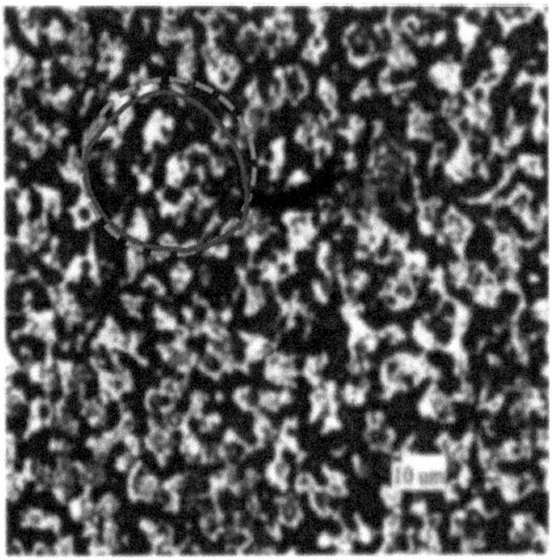

Figure 8: Active contours obtained for sample 1

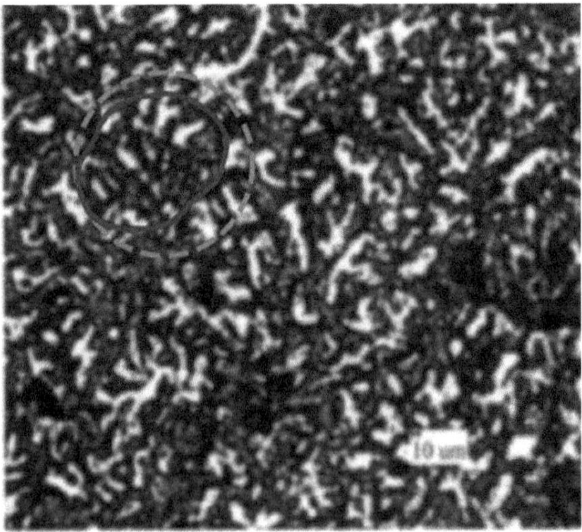

Figure 9: Active contours obtained for sample 2

Figure 10: Active contours obtained for sample 3

The improved microstructure accounts for the contour's reduced variance. In contrast to sample 2 but smoother than sample 1, sample 3 in Fig. 10 displays an intermediate behaviour with moderately coarse grains that cause somewhat wavy outlines. The outlines closely match the observed changes in the quality of the microstructure. Due to its ideal microstructure and tiny phase dispersions, sample 2 demonstrates the most reliable segmentation with smooth and clean contour convergence. Samples 1 and 3 exhibit abnormalities in the evolving contours as a result of more diffuse phase boundaries.

4. Conclusion

In this study, squeeze casting and hot extrusion were used to fabricate magnesium-based metal matrix composites reinforced with boron carbide, titanium carbide, and carbon nanotubes. Significant discrepancies between the three specimens were found during microstructural analysis, which was attributable to variances in matrix alloy composition and reinforcement type/fraction. Following sample polishing and etching, optical microscopy revealed that the best characteristics were produced by a finely tuned, homogenous microstructure with evenly dispersed fine reinforcements. In contrast, sliding faults were more common in coarse-grained microstructures with uneven phase distributions. All samples' microstructural characteristics could be clearly distinguished using image segmentation employing active contours. For the fine, uniform microstructure, the developing contours nicely tracked phase boundaries. In the evolution of contours for coarser, distributed microstructures, more irregularities appeared. The performance of active contour segmentation and qualitative microstructural variations between specimens were well associated. This demonstrates how quantitative microstructure image analysis may be used for thorough materials characterization.

On the basis of this research, future studies can investigate further metal matrix composite compositions with various matrices, reinforcements, and fractions. Image analysis and mechanical testing can be used to determine correlations between microstructure and properties. By include the proper energy terms and shape constraints, the active contour algorithm can be made to work better for composites. Annotated micrographs can also be used to train convolutional neural networks that are more sophisticated. It is possible to explore detailed 3D microstructure modelling to enable simulations of deformation behaviour. In-depth knowledge of interfaces can be obtained by the combination of image analysis and electron microscopy. For the development and optimization of composite materials for high-performance applications, microstructure segmentation and machine learning have enormous potential.

References

1. Mortensen, A. and Llorca, J. 2010. Metal matrix composites. *Annual Review of Materials Research*, 40: 243-270.
2. Rohatgi, P.K. 1993. Metal matrix composites. *Defence Science Journal*, 43(4): 323.
3. Evans, A., San Marchi, C., Mortensen, A., Evans, A., San Marchi, C. and Mortensen, A. 2003. *Metal Matrix Composites* (pp. 9-38). Springer US.
   Kainer, K.U. 2006. Basics of metal matrix composites. *In: Metal Matrix Composites: Custom-made Materials for Automotive and Aerospace Engineering*, pp. 1-54.
4. Kainer, K.U. 2006. Basics of metal matrix composites. *In: Metal Matrix Composites: Custom-made Materials for Automotive and Aerospace Engineering*, pp. 1-54.
5. Ralph, B., Yuen, H.C. and Lee, W.B. 1997. The processing of metal matrix composites – An overview. *Journal of Materials Processing Technology*, 63(1-3): 339-353.
6. Miracle, D.B. 2005. Metal matrix composites – From science to technological significance. *Composites Science and Technology*, 65(15-16): 2526-2540.
7. Clyne, T.W. and Withers, P.J. 1993. *An Introduction to Metal Matrix Composites*. Cambridge University Press.
8. Rawal, S.P. 2001. Metal-matrix composites for space applications. *Jom*, 53: 14-17.
9. Krishnan, R., Pandiaraj, S., Muthusamy, S., Panchal, H., Alsoufi, M.S., Ibrahim, A.M.M. and Elsheikh, A. 2022. Biodegradable magnesium metal matrix composites for biomedical implants: Synthesis, mechanical performance, and corrosion behaviour – A review. *Journal of Materials Research and Technology*, 20: 650-670.
10. Arora, G.S., Saxena, K.K., Mohammed, K.A., Prakash, C. and Dixit, S. 2022. Manufacturing techniques for Mg-based metal matrix composite with different reinforcements. *Crystals*, 12(7): 945.

11. Tayyebi, M. and Alizadeh, M. 2022. A novel two-step method for producing Al/Cu functionally graded metal matrix composite. *Journal of Alloys and Compounds*, 911: 165078.
12. Usca, Ü.A., Şap, S., Uzun, M., Kuntoğlu, M., Salur, E., Karabiber, A. et al. 2022. Estimation, optimization and analysis based investigation of the energy consumption in machinability of ceramic-based metal matrix composite materials. *Journal of Materials Research and Technology*, 17: 2987-2998.
13. Santhosh, M.S., Natrayan, L., Kaliappan, S., Patil, P.P., Rao, Y.S., Kumar, T.N. et al. 2022. Mechanical and wear behaviour of nano-fly ash particle-reinforced mg metal matrix composites fabricated by stir casting technique. *Journal of Nanomaterials*, 2022: 1-8.
14. Khanna, V., Singh, K., Kumar, S., Bansal, S.A., Channegowda, M., Khalid, M. and Chaudhary, V. 2022. Engineering electrical and thermal attributes of two-dimensional graphene reinforced copper/aluminium metal matrix composites for smart electronics. *ECS Journal of Solid State Science and Technology*, 11(12): 127001.
15. Mandal, V., Tripathi, P., Kumar, A., Singh, S.S. and Ramkumar, J. 2022. A study on selective laser melting (SLM) of TiC and B4C reinforced IN718 metal matrix composites (MMCs). *Journal of Alloys and Compounds*, 901: 163527.
16. Bahedh, A.S., Mishra, A., Al-Sabur, R. and Jassim, A.K. 2022. Machine learning algorithms for prediction of penetration depth and geometrical analysis of weld in friction stir spot welding process. *Metallurgical Research & Technology*, 119(3): 305.
17. Sefene, E.M., Tsegaw, A.A. and Mishra, A. 2022. Process parameter optimization of 6061AA friction stir welded joints using supervised machine learning regression-based algorithms. *Journal of Soft Computing in Civil Engineering*, 6(1): 127-137.
18. Jatti, V.S., Dhabale, R.B., Mishra, A., Khedkar, N.K., Jatti, V.S. and Jatti, A.V. 2022. Machine learning based predictive modeling of electrical discharge machining of cryo-treated NiTi, NiCu and BeCu alloys. *Applied System Innovation*, 5(6): 107.
19. Zhu, M., Wang, J., Yang, X., Zhang, Y., Zhang, L., Ren, H. et al. 2022. A review of the application of machine learning in water quality evaluation. *Eco-Environment & Health*, 1(2): 107-116.
20. Hopkins, E. 2022. Machine learning tools, algorithms, and techniques. *Journal of Self-Governance and Management Economics*, 10(1): 43-55.
21. Martins, R.M. and Gresse Von Wangenheim, C. 2023. Findings on teaching machine learning in high school: A ten-year systematic literature review. *Informatics in Education*, 22(3): 421-440.
22. Sawant, D.A., Jatti, V.S., Mishra, A. et al. 2023. Surface roughness and surface crack length prediction using supervised machine learning-based approach of electrical discharge machining of deep cryogenically treated NiTi, NiCu, and BeCu alloys. *Int J Adv Manuf Technol*, 128: 5595-5612. https://doi.org/10.1007/s00170-023-12269-1
23. Jatti, V.S., Jatti, A.V., Mishra, A., Dhabale, R.D. and Sefene, E.M. 2023. Optimizing flexural strength of fused deposition modelling using supervised machine learning algorithms. *International Journal of Information Technology*, 1-8.
24. Mishra, A., Jatti, V.S., Sefene, E.M. and Paliwal, S. 2023. Explainable artificial intelligence (XAI) and supervised machine learning-based algorithms for prediction of surface roughness of additively manufactured polylactic acid (PLA) specimens. *Applied Mechanics*, 4(2): 668-698.
25. Mishra, A. and Jatti, V.S. 2023. Novel coupled genetic algorithm–machine learning approach for predicting surface roughness in fused deposition modeling of polylactic

acid specimens. *Journal of Materials Engineering and Performance*, https://doi.org/10.1007/s11665-023-08379-2

26. Mishra, A. and Jatti, V.S. 2023. Reinforcement learning based approach for the optimization of mechanical properties of additively manufactured specimens. *International Journal on Interactive Design and Manufacturing (IJIDeM)*, 1-9.
27. Mishra, A., Jatti, V.S., Khedkar, N.K., Dhabale, R.B. and Jatti, A.V. 2023. Computer vision algorithm for the detection of fracture cracks in Oil Hardening Non-Shrinking (OHNS) die steel after machining process. *Frattura ed IntegritàStrutturale*, 17(63): 234-245.
28. Mishra, A. and Dasgupta, A. 2022. Supervised and unsupervised machine learning algorithms for forecasting the fracture location in dissimilar friction-stir-welded joints. *Forecasting*, 4(4): 787-797.
29. Mishra, A. and Morisetty, R. 2022. Determination of the Ultimate Tensile Strength (UTS) of friction stir welded similar AA6061 joints by using supervised machine learning based algorithms. *Manufacturing Letters*, 32: 83-86.
30. Dewan, M.W., Huggett, D.J., Liao, T.W., Wahab, M.A. and Okeil, A.M. 2016. Prediction of tensile strength of friction stir weld joints with adaptive neuro-fuzzy inference system (ANFIS) and neural network. *Materials & Design*, 92: 288-299.
31. Quarto, M., Bocchi, S., D'Urso, G. and Giardini, C. 2022. Hybrid finite elements method-artificial neural network approach for hardness prediction of AA6082 friction stir welded joints. *International Journal of Mechatronics and Manufacturing Systems*, 15(2-3): 149-166.
32. Sefene, E.M., Tsai, Y.H., Jamil, M., Jatti, V.S., Mishra, A., AsmareTsegaw, A. and Costa, E.C. 2023. A multi-criterion optimization of mechanical properties and sustainability performance in friction stir welding of 6061–T6 AA. *Materials Today Communications*, 36: 106838.
33. Mishra, A., Potnis, M.S., Sapre, M.S. and Jatti, V.S. 2023. Fracture analysis of friction stir spot welded acrylonitrile butadiene styrene sheet in butt configuration. *Materials Research Express*, 10(5): 055302.
34. Kim, M.G. 2020. Improved measurement of thin film thickness in spectroscopic reflectometer using convolutional neural networks. *International Journal of Precision Engineering and Manufacturing*, 21(2): 219-225.
35. Kahkeshan, B. and Hassan, S.I. 2017. Assessment of accuracy enhancement of back propagation algorithm by training the model using deep learning. *Oriental Journal of Computer Science and Technology*, 10(2).

CHAPTER

4

Experimental Investigation of Bagasse Ash in Concrete Material

Pratibha S. Agrawal[1*], Pramod N. Belkhode[1], Sarika Modak[2]

[1] Laxminarayan Innovation Technological University, Nagpur

[2] Priyadarshini College of Engineering, Nagpur

1. Introduction

Agriculture is the backbone of India and agriculture represents a large part of the economy whereas 80–90% of the population's employment is directly or indirectly based on agriculture. These agricultural products are only a significant crucial process that improves the usability of the products. Throughout the experimental program, ordinary Portland Cement of 53 grade, conforming to Indian Standard IS 12269-1987 was used. Cement must develop the appropriate strength in order to be used safely. It must represent the appropriate rheological behaviour. Generally, similar types of cement have quite different rheological and strength characteristics, particularly when used in combination with admixtures and cementing material.

Researchers all over the world are focusing on ways of utilizing either industrial or agricultural waste, as a source of raw materials for industry. This waste, utilization would not only be economical but may also result in foreign exchange earnings and environmental pollution control. A good solution to the problem of recycling agro-industrial excess would be by burning them in a controlled environment and using the ashes (waste) for more civic-minded means. Utilization of such wastes as cement and fine aggregate replacement materials may reduce the cost of concrete production and also minimize the harmful environmental effects of disposal of these wastes.

Waste obtained from industries such as slag from blast furnaces, and fly ash is generally utilized in the production of cement as supplementary materials. Agro

*Corresponding author: pratibha3674@gmail.com

waste such as sugar cane bagasse is the type of fibrous waste obtained from the sugar refining industry, along with ethanol vapour. Agro waste product such as ash obtained from sugar cane bagasse is dangerous to environmental pollution as it contains aluminium ion and silica content. India is the second largest country in the production of sugar after Brazil. Bagasse is the waste generated after the sugar production process which can be used for many applications such as the various by-products that can be used as raw material and also used for combustion purposes in the boiler for power generation. The ash obtained after burning sugar cane is known as sugar cane bagasse ash, it is the waste product of the combustion of bagasse for energy in sugar factories and is becoming an environmental burden due to the fact that it is disposed of in the landfill. Around 25.65% of bagasse (at a moisture content of 50%) and 0.61% of residual ash is produced from one ton of sugar cane [1–3].

2. Materials and Methods

In the following experiment, the material used is cement of 53 grade conforming to Indian Standard IS 12269-1987. The cement used is tested to check its appropriate strength and rheological behaviour this is because strength and rheological characteristics vary for the different types of cement. This is important to check as its combination changes with admixtures and cementing material as shown in Table 1.

Table 1: Chemical composition of cement [4, 5]

Oxide	Percentage content
CaO	60–67
SiO_2	17–25
Al_2O_3	3–8
Fe_2O_3	0.5–6
MgO	0.1–4
Alkalies (K_2O, Na_2O)	0.4–1.3
SO_3	1.3–3

2.1 Aggregates

An aggregate is a natural deposit of sand and gravel and also gives structure to the concrete. It occupies almost 75% to 80% of the volume in concrete and hence shows an influence on various properties such as workability, strength, durability and economy of concrete. To increase the density of concrete aggregate it is frequently used in different sizes. The aggregate acts as reinforcement and introduces strength to the overall composite material [6, 7]. Aggregate is also used as a base material for roads, railroads and under foundations due to its strength.

2.2 Coarse Aggregate

An aggregate having a size of more than 4.75 mm is termed as coarse aggregate. The graded coarse aggregate is described by its nominal size i.e. 40 mm, 20 mm, 16 mm, 12.5 mm etc. An 80 mm-sized aggregate is the maximum size that could be conveniently used for making concrete [8–10]. In this study coarse aggregate is conformed to IS: 383. In this experiment, the aggregate passed through a 20 mm sieve was used.

2.3 Fine Aggregate

Aggregate that passes through a 4.75 mm IS sieve and has no more than 5 per cent coarse material is known as fine aggregate. The main function of fine aggregate is to fill the voids in between coarser particles,it also helps in producing workability and uniformity in the mixture [11, 12]. In this study, fine aggregate is conformed to IS: 383.

2.4 Sugarcane Bagasse Ash

Sugarcane bagasse ash is a by-product of sugar factories and is created after burning sugarcane bagasse, which itself is created after the extraction of all economical sugar from sugarcane. The disposal of this material is already causing environmental problems around sugar factories. On the other hand, the boost in construction activities in the country created a shortage in most of the concrete making materials available, especially cement, resulting in an increase in price. This study examined the potential use of sugarcane bagasse ash as a partial cement replacement material [13].

2.5 Jute Fibre

Jute is a long, soft, shiny vegetable fibre that can be spun into coarse, strong threads. It is produced primarily from plants in the genus Corchorus, which was once classified with the family Tiliaceae and more recently with Malvaceae. The primary source of the fibre is Corchorus olitorius, but it is considered inferior to the Corchorus capsularis plant. *Corchorus* "Jute" is the name of the plant or fibre used to make burlap, hessian or gunny cloth.

Jute is one of the most affordable natural fibres available, and second only to cotton in the amount produced and variety of uses [14]. Jute fibres are composed primarily of the plant materials cellulose and lignin. The industrial term for jute fibre is *raw jute*. The fibres are off-white to brown, and 1–4 metres (3–13 feet) long. Jute is also called the *golden fibre* for its color and high cash value [15]. In this experiment we have used jute fibre of the size 10–15 mm.

2.6 Water

Water is second most important ingredient after cement, in the concrete making process. The careless use of water can lead to poor quality concrete. Therefore, a detailed study of the quantity and quality of water required for making good quality concrete is essential [16, 17]. Thus, the purpose of water in concrete is threefold. Firstly, it distributes the cement evenly, secondly, it reacts with cement chemically and produces calcium silicate hydrate gel and thirdly, it provides for the workability of concrete.

2.7 Properties of Cement

The properties of concrete materials, and the paste, mortar involved in construction materials is improved by mixing in the bagasse ash, which results in an increase in compressive strength, water tightness and fineness. Further this improvement is mainly due to presence of silica in large quantities in the bagasse ash, though this improvement varies based on the silicate contents and other properties of the raw materials. Cement production involved in the pozzolanic reaction which controls the lime in concrete due the presence of silicate. Along with sugarcane production, jute production is on large scale particularly in India. The fibres from jute are beneficial towards reducing the brittle behaviour of concrete, as jute is most reasonable of natural fibres. Jute fibre enhances the energy absorption capacity of the concrete which prevents it from failure. The investigation of the amount of jute fibres to be incorporated, so that maximum compressive strength can be sustained is yet to be identified in current research work [18]. Mixing of the natural fibre in the cement composites enhances the main benefits such as low cost, easy availability, eco-friendly, high strength, and toughness as shown in Tables 2 and 3.

Table 2: Properties of cement [19, 20]

Characteristics	Value obtained	Standard value
Fineness	3.5%	< 10
Specific gravity	3.15	–

Table 3: Chemical composition of SCBA [22, 23]

Test	Result
Moisture content	1.073%
Ash content	60%
Carbon content	38.8%
Fineness modulus	8.91
Specific gravity	2.68
Density	1.47
Volatile matter	0.127%

After burning sugarcane in the sugar factories, the ash produced is sugarcane bagasse ash. This by-product is obtained from the extraction of all economical sugar from sugarcane. It creates an environmental problem if the by-product, sugarcane bagasse ash is disposed of in landfills. This problem is overcome by using this ash in construction activities such as using it in concrete-making materials, particularly cement which solves both the problems as it boosts construction activities in the country and controls the prices of cement [21]. The main focus of the paper is to investigate sugarcane bagasse ash and jute for concrete mixture enhancement.

3. Agro Waste to Wealth

Sugarcane bagasse ash is a by-product of sugar factories found after burning sugarcane bagasse which itself is created after the extraction of all the economical sugar from sugarcane [24]. The disposal of this material is already causing environmental problems around the sugar factories. On the other hand, the boost in construction activities in the country created a shortage in most of the concrete-making materials especially cement, resulting in an increase in price [25]. This study examined the potential use of sugarcane bagasse ash as a partial cement replacement material.

The compressive strength of concrete decreases when compared to the compressive strength of concrete containing different percentages of SCBA [26]. The compressive strength of the M1 mix decreases by 8.28% after 7 days, 6.62% after 14 days and 5.88% after 28 days of curing compared with the compressive strength of concrete containing different percentages of SCBA (40 : 5). For the M2 mix, the compressive strength decreases by 17.62% after 7 days, 13.94% after 14 days and 11.38 after 28 days on comparison with the compressive strength of concrete containing a different percentage of SCBA (30 : 10). For the M3 mix, the compressive strength decreases by 23.4% after 7 days, 19.01% after 14 days and 14.53% after 28 days on comparison with the compressive strength of

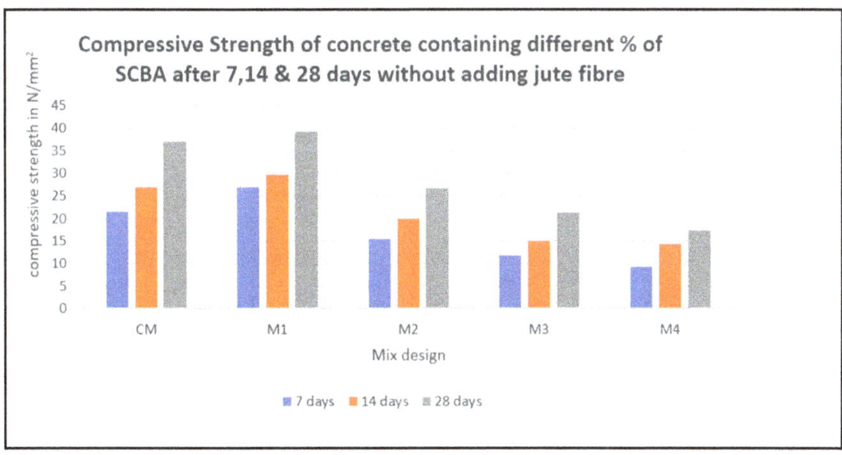

Figure 1: Compressive strength of concrete with a different percentage of SCBA after 7, 14 and 28 days

concrete containing a different percentage of SCBA (30 : 15). For the M4 mix, the compressive strength decreases by 29.63% after 7 days, 20.93% after 14 days and 18.28% after 28 days on comparison with the compressive strength of concrete containing a different % of SCBA (10 : 20). Figure 1 shows that the compressive strength decreases slightly [27].

4. Experimental Optimization of Concrete

In the dry oven small amounts of coarse aggregate is kept with the help of a pan heated to approximately a temperature of 100 to 110 degrees centigrade and a further sample weight is measured in grams. The fine nature of the coarse aggregate is measured which is termed as fineness modulus of coarse aggregates. Sieves and a mechanical shaker are used to test the fineness, in this experiment sieves are used in a descending order. The observation is recorded by properly sieving the sample weights retained on each sieve and then the recorded cumulative weight of retained particles and its percentages are recorded and calculated which helps us to find out the final fineness modulus [28, 29]. A lignite sample is passed through a 212-micron IS sieve in the dry condition, and a further lignite sample is kept in the crucible with a distribution which does not exceed 0.15 g/cm^2. This is followed by heating the crucible in the muffle furnace. This is done by raising the temperature from room temperature to 500°C in 30 min. and to 815°C in a further 30 to 60 min. After maintaining the crucible for around 50 to 60 minutes it is taken out to be cooled and allowed to cool first on a cold metal slab for 10 minutes, and finally in a desiccator for 15 minutes. The crucible weight is measured at an empty level and after taking it out from the furnace [30, 31]. Further re-ignition is carried out at the same temperature until the change in mass of the ash is less than 0.001 g. Afterwards the colour of the ash was observed and brushed out. The disc or crucible was reweighted and ash percentages were obtained by using the following formula:

Ash percentages,

$$A = 100 \times (M3–M4)/(M2–M1)$$

where,
M1 = mass in gram of dish
M2 = mass in gram of dish and sample
M3 = mass in gram of dish and ash
M4 = mass in gram of dish after brushing out the ash and on reweighing

5. Results and Discussion

5.1 Compression Test Result (IS 516-1959)

Table 4 and Fig. 2 show that the compressive strength of concrete decreases when compared to the compressive strength of concrete containing different percentages of SCBA and a different percentage of jute fibre. The compressive strength of the M1 mix decreases by 8.28% after 7 days, 6.62% after 14 days and 5.88% after 28 days of curing, compared with the compressive strength of concrete containing different percentages of SCBA [32, 33].

Table 4: Compressive strength of concrete containing different percentage of SCBA after 7, 14 and 28 days

Mix (Cement : SCBA)	Avg. compressive strength (N/mm^2)		
	7 Days	**14 Days**	**28 Days**
CM (100 : 0)	21.44	26.96	37.04
M1 (40 : 5)	26.90	29.60	39.3
M2 (30 : 10)	15.56	18.93	26.81
M3 (20 : 15)	13.78	16.67	21.94
M4 (10 : 20)	9.33	14.15	17.4

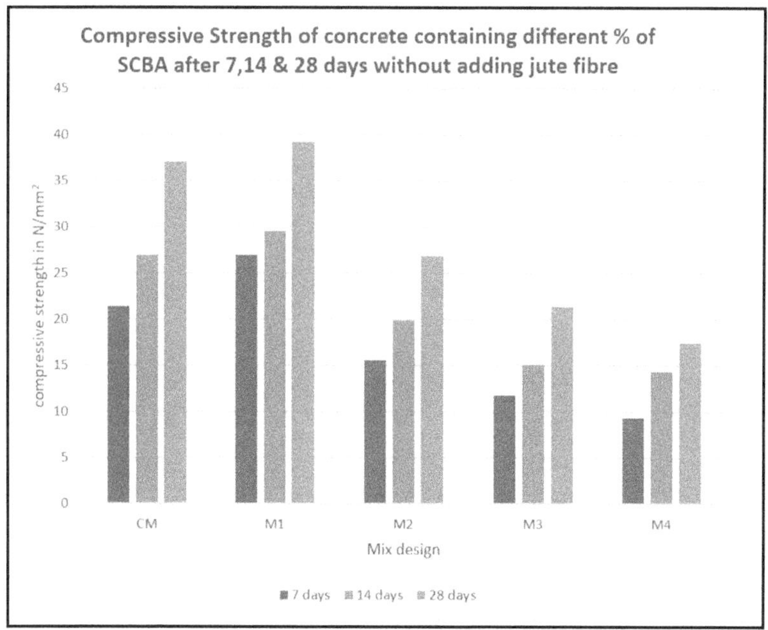

Figure 2: Compressive strength of concrete with different percentage of SCBA after 7, 14 and 28 days

For the M2 mix, the compressive strength decreased by 17.62% after 7 days, 13.94% after 14 days and 11.38 after 28 days compared with the compressive strength of concrete containing a different % of SCBA. For the M3 mix, the compressive strength decreased by 23.4% after 7 days, 19.01% after 14 days and 14.53% after 28 days compared with the compressive strength of concrete containing different % of SCBA. For the M4 mix, the compressive strength decreased by 29.63% after 7 days, 20.93% after 14 days and 18.28% after 28 days compared with the compressive strength of concrete containing a different percentage of SCBA. This shows that the compressive strength slightly decreases [34].

5.2 Artificial Neural Network (ANN) Simulation

Artificial Neural Network (ANN) simulation is concerned with a multilayer feed-forward topology for the network. MATLAB software is used for training the network for response variables and to calculate percentage error plots for prediction for the network for the concerned variables [35]. Comparison graphs of experimental data and artificial neural network responses for each of the dependent variables [36]. These graphs proved to be successful in terms of agreement with actual values of observations and neural output. The simulation consists of three layers. The first layer is known as the input layer. The five independent variables are equal to the five neurons in the input layer. The hidden layer is the second layer. The output layer is the third layer. It contains one neuron as one of the dependent variables of compressive strength [37].

MATLAB software is selected for developing ANN. The program executed for compressive strength is as follows

```
clear all;
close all;
inputs3=[
]
a1=inputs3
a2=a1
input_data=a2;
Output3= [
]
y1=output3
y2=y1
size(a2);
size(y2);
p=a2';
sizep=size(p);
t=y2';
sizet=size(t);
[S Q] =size(t)
[pn,meanp,stdp,tn,meant,stdt] = prestd(p,t);
```

```
net = newff(minmax(pn), [10 1], {'logsig' 'purelin'},'trainlm');
net.performFcn='mse';
net.trainParam.goal=.99;
net.trainParam.show=200;
net.trainParam.epochs=50;
net.trainParam.mc=0.05 ;
net = train(net,pn,tn);
an = sim(net,pn);
[a] = poststd(an,meant,stdt);
error=t-a;
x1=1:16;
plot (x1, t,'rs-', x1, a,'b-')
legend('Experimental','Neural');
title ('Output (Red) and Neural Network Prediction (Blue) Plot');
xlabel('Experiment No.');
ylabel('Output');
grid on;
figure
error_percentage=100*error. /t
plot (x1, error_percentage)
legend ('percentage error');
axis ([0 10 -100 100]);
title ('Percentage Error Plot in Neural Network Prediction');
xlabel('Experiment No.');
ylabel('Error in %');
grid on;
for ii=1:16
xx1=input_data(ii,1);
yy2=input_data(ii,2);
zz3=input_data(ii,3);
xx4=input_data(ii,4);
pause
yyy(1,ii)=power(xx1,0.1)*power(yy2,0.09)*power(zz3,-0.2)* power(xx4,-0.07);
yy_practical(ii)=(y2(ii,1));
yy_eqn(ii)= (yyy (1, ii))
yy_neur(ii)= (a (1, ii))
yy_practical_abs(ii)=(y2(ii,1));
yy_eqn_abs(ii)= (yyy (1, ii));
yy_neur_abs(ii)= (a (1, ii));
pause
end
figure;
```

```
plot (x1, yy_practical_abs,'r-', x1, yy_eqn_abs,'b-', x1, yy_neur_abs,'k-');
legend('Practical','Equation','Neural');
title ('Comparision between practical data, equation-based data and neural based data');
xlabel('Experimental');
grid on;
figure;
plot (x1, yy_practical_abs,'r-', x1, yy_eqn_abs,'b-');
legend ('Practical','Equation');
title ('Comparison between practical data, equation-based data and neural based data');
xlabel('Experimental');
grid on;
figure;
plot (x1, yy_practical_abs,'r-', x1, yy_neur_abs,'k-');
legend('Practical','Neural');
title ('Comparison between practical data, equation-based data and neural based data');
xlabel('Experimental');
grid on;
error1=yy_practical_abs-yy_eqn_abs
figure
error_percentage1=100*error1. /yy_practical_abs;
plot (x1, error_percentage,'k-', x1, error_percentage1,'b-');
legend('Neural','Equation');
axis ([0 16 -100 100]);
title ('Percentage Error Plot in Equation (blue), Neural Network (black) Prediction');
xlabel('Experiment No.');
ylabel('Error in %');
grid on;
meanexp=mean(output3)
meanann=mean(a)
meanmath=mean(yy_eqn_abs)
mean_absolute_error_performance_function = mae(error)
mean_squared_error_performance_function = mse(error)
```

5.3 Formulation of Mathematical Model

The compressive strength of concrete is the dependent term which depends on the contents of bagasse ash in different percentages of SCBA and different percentages of jute fibre. The dependent and independent terms in the preparation of mixture containing different percentages of SCBA and different percentages of jute fibre as shown in Table 5.

Table 5: Dependent and independent variables

Sr. No.	Terms	Nature of variables
01	Compressive strength	Dependent
02	Mixture M1	Independent
03	Mixture M2	Independent
04	Mixture M3	Independent
05	Mixture M4	Independent
06	Mixture M5	Independent

Compressive Strength = K $(M1)^a (M2)^b (M3)^c (M4)^d (M5)^e$
Where, K = Constant, considering the influence of remaining variables

From the above comparison of the phenomenal response by a conventional approach and ANN simulation, it seems to be that the curves obtained are overlapping due to the lower percentage of error which is on the positive side (Figs. 3 and 4).

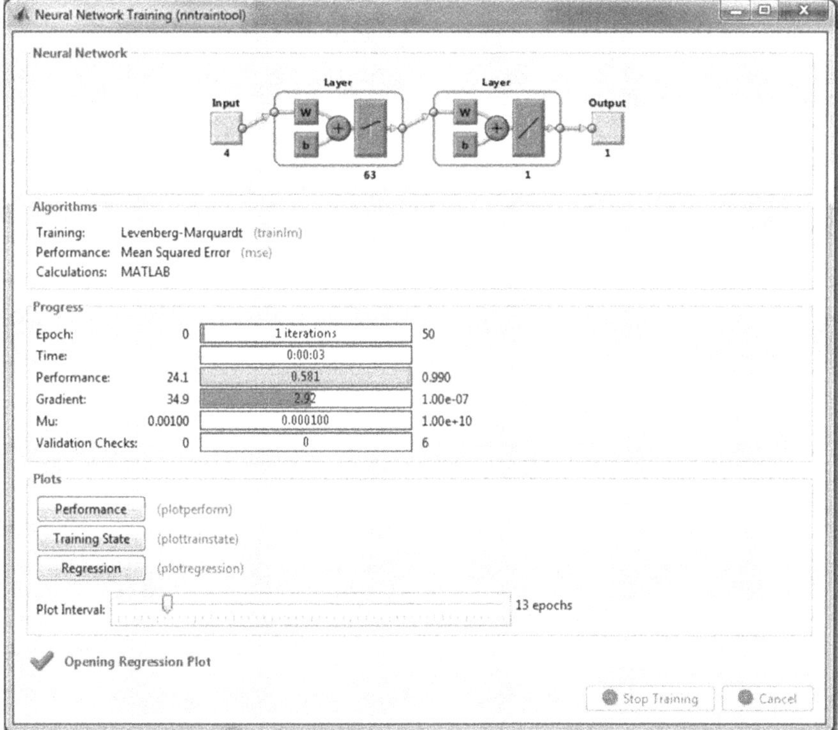

Figure 3: Training of the network for compressive strength

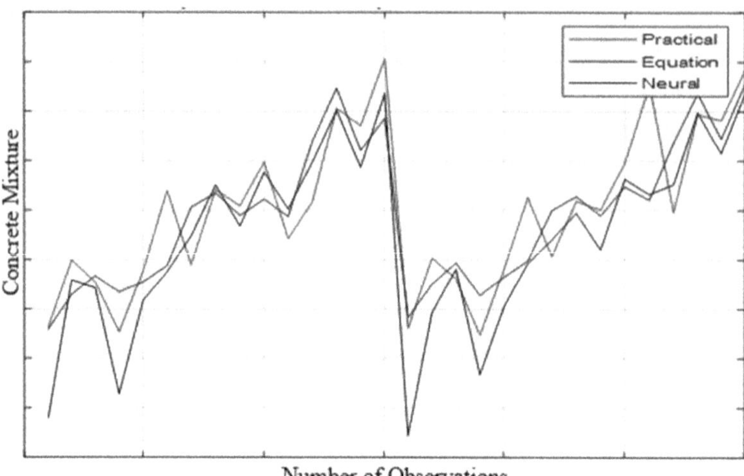

Figure 4: Graph of comparison with experimental database and neural network prediction for compressive strength

6. Conclusion

The compressive strength of the concrete increased at a 5% replacement of cement by SCBA and 0.1% jute fibre and then gradually decreased. The maximum compressive strength was obtained at a 5% replacement of cement by SCBA and 0.1% jute fibre as 29.33% at 7 days, 31.7% at 14 days and 41.63% at 28 days. The strength has been increased by 11.02% as compared to CM. The compressive strength of the concrete increases as the percentage of jute fibre increases till M4 and then decreases.

References

1. Vijay Vikram, A.S. and Arivalagan, S. 2017. Engineering properties on the sugar cane bagasse with sisal fibre reinforced concrete. *Int. J. Appl. Eng. Res.*, 12(24): 15142-15146.
2. Xu, Q., Ji, T., Gao, S.J., Yang, Z. and Wu, N. 2018. Characteristics and applications of sugar cane bagasse ash waste in cementitious materials. *Materials*, doi: 10.3390/ma12010039.
3. Berenguer, R.A., Capraro, A.P.B., Farias de Medeiros, M.H., Carneiro, A.M.P. and de Oliveira, R.A. 2020. Sugar cane bagasse ash as a partial substitute of Portland cement: Effect on mechanical properties and emission of carbon dioxide. *J. Environ. Chem. Eng.*, doi: 10.1016/j.jece.2020.103655.
4. Dasari, K.B., Sun, G.M. and Cho, H. 2021. Assessment of radiological hazards and chemical composition of cement produced in South Korea. *J. Radioanal. Nucl. Chem.*, doi: 10.1007/s10967-021-07936-0.

5. Eric Opoku, A. 2013. Chemical Composition of Cement, *Adv. Mater. Sci.*, 2015: 1-5.
6. Ouda, A.S. 2015. Development of high-performance heavy density concrete using different aggregates for gamma-ray shielding. *Prog. Nucl. Energy*, doi: 10.1016/j.pnucene.2014.11.009.
7. Mahmood, A.H., Foster, S.J. and Castel, A. 2020. Development of high-density geopolymer concrete with steel furnace slag aggregate for coastal protection structures. *Constr. Build. Mater.*, doi: 10.1016/j.conbuildmat.2020.118681.
8. Wu, Y., Qi, Z., Niu, M., Yao, Y., Luo, Z. and Zhang, K. 2022. Effect of moisture condition of brick–concrete recycled coarse aggregate on the properties of concrete. *Materials (Basel)*, doi: 10.3390/ma15207204.
9. Yuan, Y., Shao, Z., Qiao, R., Guo, X. and Wang, W. 2022. Crack damage evolution in concrete coarse aggregates under microwave-induced thermal stress. *Arch. Civ. Mech. Eng.*, doi: 10.1007/s43452–022–00419–3.
10. Azúa, G., González, M., Arroyo, P. and Kurama, Y. 2019. Recycled coarse aggregates from precast plant and building demolitions: Environmental and economic modeling through stochastic simulations. *J. Clean. Prod.*, doi: 10.1016/j.jclepro.2018.11.049.
11. Santha Kumar, G. and Minocha, A.K. 2018. Studies on thermo-chemical treatment of recycled concrete fine aggregates for use in concrete. *J. Mater. Cycles Waste Manag.*, doi: 10.1007/s10163–017–0604–6.
12. Nedeljković, M., Visser, J., Šavija, B., Valcke, S. and Schlangen, E. 2021. Use of fine recycled concrete aggregates in concrete: A critical review. *Journal of Building Engineering*, doi: 10.1016/j.jobe.2021.102196.
13. Dhengare, S., Amrodiya, S., Shelote, M., Asati, A., Bandwal, N., Khangan, A. and Jichkar, R. 2015. Utilization of sugarcane bagasse ash as a supplementary cementitious material in concrete and mortar – A review.
14. Chakraborty, S., Kundu, S.P., Roy, A., Basak, R.K., Adhikari, B. and Majumder, S.B. 2013. Improvement of the mechanical properties of jute fibre reinforced cement mortar: A statistical approach. *Constr. Build. Mater.*, doi: 10.1016/j.conbuildmat.2012.09.067.
15. Wang, H., Memon, H., Hassan, E.A.M., Miah, M.S. and Ali, M.A. 2019. Effect of jute fiber modification on mechanical properties of jute fiber composite. *Materials (Basel)*, doi: 10.3390/ma12081226.
16. Wang, F., Li, K. and Liu, Y. 2022. Optimal water-cement ratio of cement-stabilized soil. *Constr. Build. Mater.*, doi: 10.1016/j.conbuildmat.2021.126211.
17. Mehdizadeh, H., Jia, X., Mo, K.H. and Ling, T.C. 2021. Effect of water-to-cement ratio induced hydration on the accelerated carbonation of cement pastes. *Environ. Pollut.*, doi: 10.1016/j.envpol.2021.116914.
18. Sorsa, A. 2022. Engineering properties of cement stabilized expansive clay soil. *Civ. Environ. Eng.*, doi: 10.2478/cee-2022–0031.
19. Horszczaruk, E. 2019. Properties of cement-based composites modified with magnetite nanoparticles: A review. *Materials*, doi: 10.3390/ma12020326.
20. Krishnya, S., Yoda, Y. and Elakneswaran, Y. 2021. A two-stage model for the prediction of mechanical properties of cement paste. *Cem. Concr. Compos.*, doi: 10.1016/j.cemconcomp.2020.103853.
21. Athira, G. and Bahurudeen, A. 2022. Rheological properties of cement paste blended with sugarcane bagasse ash and rice straw ash. *Constr. Build. Mater.*, doi: 10.1016/j.conbuildmat.2022.127377.
22. Prabhath, N., Kumara, B., Vithanage, V., Indupama, A., Sewwandi, N., Maduwantha, K. et al. 2022. A review on the optimization of the mechanical properties of sugarcane-

bagasse-ash-integrated concretes. *Journal of Composites Science*, doi: 10.3390/jcs6100283.
23. Chuewangkam, N., Nachaithong, T., Chanlek, N., Thongbai, P. and Pinitsoontorn, S. 2022. Mechanical and dielectric properties of fly ash geopolymer/sugarcane bagasse ash composites. *Polymers (Basel)*, doi: 10.3390/polym14061140.
24. Khawaja, S.A., Javed, U., Zafar, T., Riaz, M., Zafar, M.S. and Khan, M.K. 2021. Eco-friendly incorporation of sugarcane bagasse ash as partial replacement of sand in foam concrete. *Clean. Eng. Technol.*, doi: 10.1016/j.clet.2021.100164.
25. Javed, F., Amin, N., Shah, I., Khan, K., Iftikhar, B., Farooq, F. et al. 2020. Applications of gene expression programming and regression techniques for estimating compressive strength of bagasse ash based concrete. *Crystals*, doi: 10.3390/cryst10090737.
26. Seyoum, R., Tesfamariam, B.B., Andoshe, D.M., Algahtani, A., Ahmed, G.M.S. and Tirth, V. 2021. Investigation on control burned of bagasse ash on the properties of bagasse ash-blended mortars. *Materials (Basel)*, doi: 10.3390/ma14174991.
27. Tarekegn, M., Getachew, K. and Kenea, G. 2022. Experimental investigation of concrete characteristics strength with partial replacement of cement by hybrid coffee husk and sugarcane bagasse ash. *Adv. Mater. Sci. Eng.*, doi: 10.1155/2022/5363766.
28. Adajar, M.A.Q. and Valbuena, K.R. 2021. Optimization of the strength properties of expansive soil stabilized with agricultural wastes. *Int. J. GEOMATE*, doi: 10.21660/2021.88.gxi253.
29. Mohan, R., Athira, G., Mali, A.K., Bahurudeen, A. and Nanthagopalan, P. 2021. Systematic pretreatment process and optimization of sugarcane bagasse ash dosage for use in cement-based products. *J. Mater. Civ. Eng.*, doi: 10.1061/(asce)mt.1943-5533.0003650.
30. Satheesh, T. and Karthikeyan, G. 2020. Experimental study on partial replacement of cement with metakaolin and bagasse ash for performance improvement. *Int. Res. J. Eng. Technol.*, 7(2): 763-768.
31. Muharja, M., Darmayanti, R.F., Widjaja, A., Manurung, Y.H., Alamsyah, I. and Fadilah, S.N. 2022. Optimization of sugarcane bagasse ash utilization for concrete bricks production using Plackett-Burman and central composite design. *J. Tek. Kim. dan Lingkung.*, doi: 10.33795/jtkl.v6i1.282.
32. Berenguer, R.A., Helene, P., Silva, F.A.N., Torres, S.M., Monteiro, E.C.B., de, A.A. and Neto, M. 2018. On the influence of sugarcane bagasse ashes as a partial replacement of cement in compressive strength of mortars. *Rev. ALCONPAT*, doi: 10.21041/ra.v8i1.187.
33. Ali, S., Kumar, A., Rizvi, S.H., Ali, M. and Ahmed, I. 2020. Effect of sugarcane bagasse ash as partial cement replacement on the compressive strength of concrete. *Quaid-e-Awam Univ. Res. J. Eng. Sci. Technol.*, doi: 10.52584/qrj.1802.07.
34. Quedou, P.G., Wirquin, E. and Bokhoree, C. 2021. Sustainable concrete: Potency of sugarcane bagasse ash as a cementitious material in the construction industry. *Case Stud. Constr. Mater.*, doi: 10.1016/j.cscm.2021.e00545.
35. Belkhode, P., Modak, J.P., Vidyasagar, V. and Agrawal, P. 2021. Analysis of performance of the model. *Mathematical Modeling and Simulation*, doi: 10.1201/9781003132127–9.
36. Chawla, K.K. 1998. *Composite Materials Science and Engineering*. Second Edition. Springer-Verlag, New York.
37. Linge, S. and Langtangen, H.P. 2016. Programming for Computations – Matlab: A gentle introduction to numerical simulations with Matlab. *in Computational Science and Engineering*. Springer Heidelberg Dordrecht, London, New York.

CHAPTER

5

Computational Material Science for Cheminformatics Feature Descriptive Language (CFDL) with Categorical Data

Chandrasekaran S.[1], Lakshmi D.[2*], Priya M.[3], Neelu Nagpal[4]

[1] Bharathiar University Coimbatore, Vadavalli, Coimbatore, Tamil Nadu – 641046
[2] School of Computing Science and Engineering, VIT Bhopal University, India
[3] Physics, Saveetha Engineering College, Chennai, India
[4] EEE Department, Maharaja Agrasen Institute of Technology, Delhi, India

1. Introduction

Hansch (1969) introduced the concept that biological activity could be related to molecular structure through the concentration of its compounds. At the same time, the quantum phenomena will not only affect its own structure and energy level dynamically but also get radiated into outer space. Specific materials have their own radiation levels and irradiations are susceptible to biological systems and human biological systems are vulnerable to these quantum-level unseen phenomena. These reactions can be referred to as quantum-level correlations or causations due to the chemical composition and decomposition processes. These causal interactions, such as photons moving from one region to another, correlate measurements taken at one observing location with measurements taken later in another observing location. For gravity to be causal indefiniteness, the entire globe does not need to be in superposition, but even a single atom, when in a superposition of two positions, defines the metric in two ways at the same time [1]. Nanoscience, correlated electrons in solids, excited electronic states, defects in solids, magnetic spin systems, and biomimetic materials are some of the emerging areas where computational material science is focusing [2]. The

*Corresponding author: lakshmi.lifefordivine@gmail.com

other area of computational materials is pointed towards quantum chemistry where quantum states spaces and transitions are discussed across multiple reactions. Many reactions in quantum chemistry have been defined from various perspectives as spin-forbidden reactions that are prevented from occurring due to a necessary transition between two quantum states. When the reactant is in one spin state and the product is in a different spin state, the related reaction will have a higher activation energy than a reaction where the reactant and product spin states are isomorphic. A lower rate of reaction is observed because of the increased activation energy [3]. Modern computational materials science has evolved into an interdisciplinary discipline that incorporates and expands on the rich portfolio of quantum approaches created in biology, chemistry, mathematics, and physics [5]. The three major sections of computational material science focusing on data aspects are material discovery, material phenomena and advanced material modeling [6]. The next area of correlation with the proposed work is in the domain of biomaterials. Biomaterials is a branch of tissue engineering that combines discoveries from biochemistry, cell and molecular biology, and materials science to create three-dimensional structures that can be used to replace or repair biological components that are damaged, missing, or not working properly. Another issue addressed is to find a common descriptive language about the new material identified or synthesized and share the information about those materials among various repositories and APIs for Material Cloud [7]. The next phase of computational material science in material discovery and phenomena is the application of machine learning techniques. In materials science, machine learning is predominantly concerned with supervised learning, as the performance of such methods is largely determined by the amount and quality of data provided. However, with recent developments in matter, energy, and quantum behaviour and dynamics, quantum-based machine learning has emerged as one of the most difficult tasks in material informatics [8]. Many works on quantum simulations and analysis are being carried out in the computational material science field. The effect of exchange-correlation interactions of the surroundings on active regions is included in a quantum embedding theory based on DFT that is scalable to large systems [9]. Humans and society are trying not only to identify new materials but also to study their quantum excitations to be useful to society, especially in healthcare investments. The new materials and their quantum behaviour are to be understood as superconductivity, super magnetism; superfluidity and quantum spin liquidity. The capability to create and control quantum materials will be a true materials revolution, with transformative effects on energy production and use, transportation, and information technology. There is a worldwide effort to realize the potential of these materials [10].

In the twenty-first century, the specific field of study known as "quantum materials" emerges, where computations and information with data science bring revolution with biological sensing and bioenergy. Quantum materials in the twenty-first century, like semiconductors in the twentieth century, will have the ability to revolutionize information, sensing, energy, and associated technologies

[11]. It has been proven from ancient times onwards that the combination of certain materials and the fusion processes of these will have some extraordinary emergent properties. The six most significant neurotransmitters used to characterize depression, schizophrenia, psychosis, the influence of nicotine or alcohol intake, alcohol dependency, delirium, drug addiction, detoxification, epilepsy, and more manage human well-being and happiness or melancholy. Because radioactive materials, like other decays, have a pure quantum effect with only a finite quantum of action, it is possible for a system to remain unchanged until it decays unexpectedly. To discover the new material and study its behavior, a technique called feature engineering is the most suitable ML model as a supervised learning strategy. This feature engineering will be machine-readable or vector of vectors or vectors of numbers to define the target properties of the materials. Once a set of features to represent a data set has been gathered, it is typical to choose a representative set of features that is large enough to produce low model errors while also avoiding model under- fitting. They must not be so massive that they result in overfitting, which reduces overall model accuracy and extrapolative capabilities [12]. Converting feature values and their class labels to a fixed-size feature vector form that can be fed into feature material engineering classifiers [13] is a fundamental problem in generalizing across diverse datasets. It is also more important to consider the domain-specific feature (DSF) suitable for the classification of the generated data set in material science. The importance of all the features of the materials is to be used in the material selection and discovery processes as shown in Fig. 1. The external appearances of these materials used for idols or sculptures in India along with their physical, atomic, and thermal properties are used to manufacture many deity idols in the country. As illustrated in Fig. 2, the features at different levels must be considered while developing a novel material or studying the material's behaviour under normal and soaked conditions. The quantum effect of these minerals can be thought of as a mix of chemical, biological, and structural space properties of different materials. We

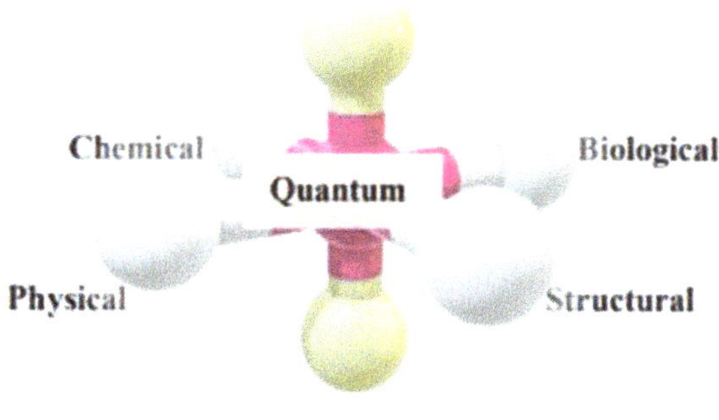

Figure 1: Multiple features with multiple levels in material discovery

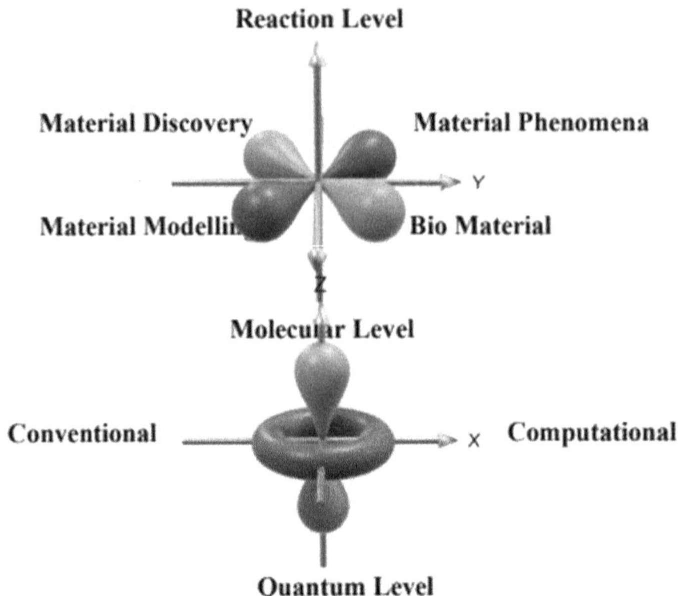

Figure 2: The importance of quantum level machine learning for material science

have identified three major research objectives for which we think data-driven approaches have the largest impact on materials science: materials discovery, understanding materials phenomena, and advancing materials modeling.

Spin Quantum Numbers are considered to be a quadruple with (n, l, m_l, m_S) which are integers representing the numbers of orbits, numbers of spaces, the magnetic spin values and spin values respectively in Table 1.

The features and the values of specific features or their attribute values of all materials can be collected. However, the challenge lies in the identification of the correct feature or characteristic which will be dominating among all other features while processing. The chemical and quantum relations across the features of all the materials have to be fully explored if the correct mixture or compound is to be discovered. In the advanced material science area, massless materials, spineless actions and stateless measurements are also more important than pure simple chemical reactions with or without catalysts. The spin or magnetic properties may be the cause for certain phenomena in the combination of materials including quantum-level gravity and causality effects.Nanomaterials' unusual size, shape, chemical content, surface structure, charge, solubility, and aggregation strongly influence their interactions with biomolecules and cells due to their unique physicochemical and biological features [20].

Quantum Gravity and Causal Effect (QGCE): The general theory of relativity introduced the concepts of conventional gravity and space, as well as the approach of time and its divergence at the black hole's centre. The quantum gravity or

Table 1: Quantum electro magnetic mass properties of Au, Ag, Fe, Cu, Sn and Zn

Sl. no.	Name	Silver	Gold	Iron	Zinc	Tin	Lead	Rhodium	Copper
1	Symbol	Ag	Au	Fe	Zn	Sn	Pb	Rh	Cu
2	Electronegativity	1.93	2.54	1.83	1.65	1.96	2.33	2.28	1.9
3	Electron Affinity	125.6 kJ/mol	222.8 kJ/mol	15.7 kJ/mol	0 kJ/mol	107.3 kJ/mol	35.1 kJ/mol	109.7 kJ/mol	118.4 kJ/mol
4	Space Group Number	225	225	229	194	141	225	225	225
5	Quantum Numbers	2S1/2	2S1/2	5D4	1S0	3P0	3P0	4F9/2	2S1/2
6	Neutron Cross Section	63.6	98.7	2.56	1.1	0.62	0.171	145	3.78
7	Neutron Mass Absorption	0.02	0.017	0.0015	0.00055	0.0002	0.00003	0.063	0.0021
8	Valence Electrons	1	5	3	2	4	4	6	2

graviton particle is interpreted when the general theory of relativity is brought to the quantum level. The Weinberg-Witten theorem will be used for massless particles with spin in most theoretical physics. According to the theorem, massless particles with spin $j > 1/2$ cannot carry a Lorentz-covariant current, and massless particles with spin $j > 1$ cannot carry a Lorentz-covariant stress energy [2]. The theorem is commonly taken to suggest that in a relativistic quantum field theory, the graviton ($j = 2$) cannot be a composite particle. The gravitational field's invisible quantum behaviour aims to combine various theories, such as string theory, into a description of everything, including electromagnetic theory. Professor Lucien Hardy, who works on quantum gravity, pioneered the study of indefinite causality as a means of comprehending the quantum nature of gravity; however, he believes that quantum gravity will inevitably inherit general relativity's radical feature of lack of fixed time and fixed causality [1]. In a similar vein to Hardy's causality approach, the Vienna researchers established a method of storing probabilities for detecting different outcomes in different locations without reference to a set background time. The Markup Feature Generation and Measured Classification yield a set of features of various minerals especially used in the fabrications of Idols collected from the literary texts available in palm leaves and copper plates. The materials have different features during different processes and in different conditions such as temperature and reactions. Dynamic feature identification and selection of monitored feature categories with all supplementary materials are to be generated as an active mandatory dynamic feature set (AMDFS) as shown in Fig. 3.

Computationally these features are to be selected based on their basic functionality, supplementary, complement ability, reactivity and timeliness from different domains of actions and reactions of the materials in the compound. The

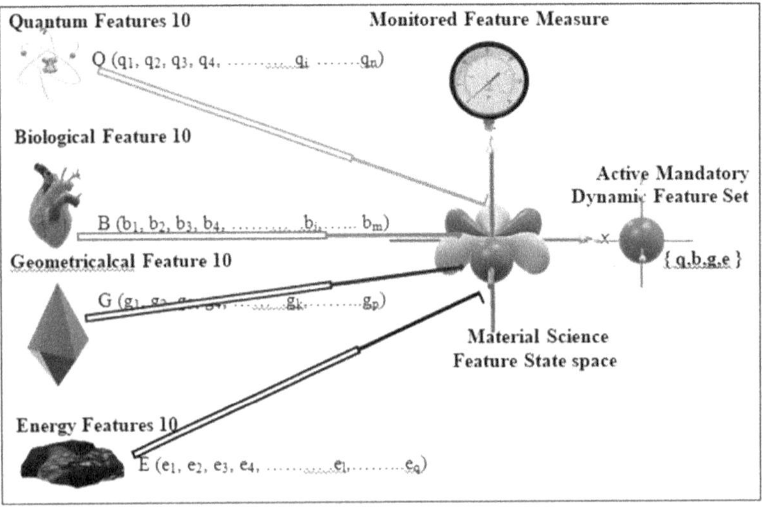

Figure 3: Multiple domain measured classification markup feature (MDMCMF)

quantum, biological, geometrical and energy features of the materials can be declared as a series of parameters and their values. For example,

Quantum features (quantum numbers, space group number, electronegativity, electron affinity, neutron mass, valency number, ...)

and

Geometrical features (crystalline, powder, reflective, structured, unstructured, unpurified, extrinsic materials, doped, ...) and so on

These sets of features of different categories are represented as the following:

$$Q\,(q_1, q_2, q_3, q_4, \dots\dots\dots\; q_i \dots\; q_n)$$
$$B\,(b_1, b_2, b_3, b_4, \dots\dots\dots .b_j \dots\; b_m)$$
$$G\,(g_1, g_2, g_3, g_4, \dots\dots\dots\; .g_k \dots\; g_p)$$
$$E\,(e_1, e_2, e_3, e_4, \dots\dots\dots .e_l \dots\; e_q)$$

A data instance's feature values operate as coalition members. In material science, the domain-specific feature of that material in that process and compound for the specific target will be called a domain-specific material feature, $\text{DSMF}_{\text{process, compound, target}}$ as shown in Fig. 4. The feature sets from the materials are collected from the data tables under various manufacturing or fusion processes. The feature values are obtained from the experimental and statistical data from multiple experiments over many areas of research.

$$f(x) = \beta_0 + \beta_1 x_1 + \dots + \beta_p x_p \tag{1}$$

The contribution in the material compound or coalition φ_j,

$$\varphi_j(f) = \beta_j x_j - E(\beta_j x_j) = \beta_j x_j - \beta_j E(x_j)$$

where $E(\beta_j x_j)$ is the mean effect estimate for feature j and where x is the instance for which it is expected to compute the contributions in the coalition or compound. Each x_j is a feature value, with $j = 1, \dots, p$. The β_j is the weight corresponding to feature j.

The features may be in controlled environment with many supplementary minerals.

The negative sign indicates the feature or parameter is to be avoided or not added as in (1).

$$\begin{bmatrix} q1 & b1 \\ g2 & e3 \end{bmatrix}\begin{bmatrix} -q4 & b6 \\ -g1 & e2 \end{bmatrix}\begin{bmatrix} q2 & -b4 \\ g5 & e1 \end{bmatrix}[25\,\text{deg}] \bigcup_{\text{supl.max}}^{\text{supl.max}} b10{\cdot}e4(\text{held}10\text{ degree})$$

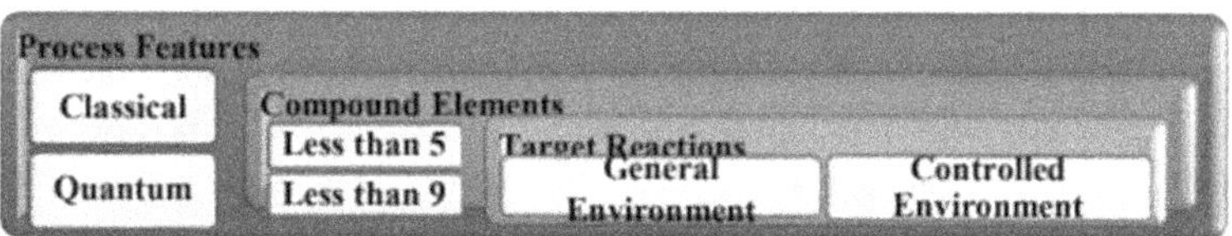

Figure 4: Domain-specific material feature, $\text{DSMF}_{\text{process, compound, target}}$

The supporting materials and the ambient temperature in which certain features will be exhibited are collected as supplementary features of the materials. Some of the features will negate or transform the other materials' features in the opposite direction. Some features cannot be visualized with the naked eye as radioactive decay and some other features when combined with other materials may turn poisonous towards vulnerable human cells. Hence the feature table is a set of values either integers or real numbers during various thermal, ecological and mechanical stresses. The features are:

- ***Feature*** *(functionality, supplementary, complement, radioactivity, vulnerability, action...)*
- ***processes*** *(heating, overheating, melting, carbonizing, cooling, mixing, proportionate mix...)*
- ***complementing*** *(increasing, decreasing, adding, diluting, concentrating, powdering...)*

The basic goal of feature fusion is to reduce the number of redundant noisy features while maintaining correct measured feature identification. It can combine two or more features from distinct domains, reducing the problem of the "curse of dimensionality". The feature selection process is at the heart of feature fusion [16]. Methods of feature selection, rating, and feature combining are increasingly important in feature fusion to produce the most predicted results. Because there are more alternatives and chances for the optimized output in both cases of fusion, it was difficult to determine the best or best collection of approaches for a particular task. The performance is determined by the dataset, the application domain, as well as the number of standards and rules.

SHADE of SHAP is a game theoretic strategy for describing the output of any machine learning model (SHapley Additive exPlanations). It connects optimal credit allocation with local explanations using standard Shapley values from game theory and associated expansions [14]. The SHAP's key objective is to compute the contribution of each feature to the prediction of every instance y in order to explain the prediction. Shapley values are computed using the SHAP explanation technique, which is based on coalitional game theory. In the same way, adding elements to materials to produce a targeted reaction of material with specific known attributes works. The instances are weighted by LIME based on their proximity to the originating instance. The weight in LIME is proportional to the number of 0s in the coalition vector. SHAP weights the sampled instances based on the coalition's Shapley value estimation weight. With a mapping between interpretable inputs and the original input space, the LIME, locally interpretable model explanation predicts a feature and its characteristics that strictly obey SHAP values [15]. The Shapley value is the sum of all marginal contributions made by all potential coalitions. The number of features increases exponentially with the number of features, thus it's best to keep the number of possible combinations in the game or chemical composition to a minimum to keep the computation time under control.

Let f be the original prediction model to be explained and g the explanation model. Here, we focus on local methods designed to explain a prediction $f(x)$ based on a single input x, as proposed in LIME. Explanation models often use simplified inputs x' that map to the original inputs through a mapping function $x = h_x(x')$. Local methods try to ensure $g(z') \approx f(h_x(z'))$ whenever $z' \approx x'$. Here it is to be understood that $h_x(x') = x$ even though x' may contain less information than x because h_x is specific to the current input x.

$$g(z') = \phi_0 + \sum_{i=1}^{M} \phi i z' i \tag{2}$$

where $z' \in \{0, 1\}M$, M is the number of simplified input features, and $\varphi_i \in R$. Methods with explanation models matching the above equation, attribute an effect φ_i to each feature, and summing the effects of all feature attributions, approximates the output $f(x)$ of the original model. The hierarchy or order in which the additive features are well explained with causal reasoning and its final values. The proposed technique starts with Shapy values with hierarchical additive domain expected feature (SHADEF) values of materials to meet the challenges of the feature engineering in the computational material science issues. The proposed SHADE techniques are not only a feature selection process but also a feature fusion where Shapley additive domain explanations are hierarchically considered, process p_i of chemical domain has to be considered first, gold with one in two million atoms doped with silver and copper without barium and tin. It is a domain specific feature engineering process to obtain a material or idol of expected emergent features with a mixture of known elements. The feature engineering phases in which hierarchical selection, absolute measurement, additive assessment and fusion explanation are carried out as explored below:

Feature Hierarchical Selection (FHS)

The process name p_i may have many attributes and their values like represented below:

p_i *(domain_name, weight_doped, additive elements, negated elements)* $= p_i (d_i, w_i, a_i, n_i)$

p_j *(domain_name, weight_doped, additive elements, negated elements)* $= p_j (d_j, w_j, a_j, n_j)$

$p_i \rightarrow p_j \rightarrow p_k \ldots\ldots \rightarrow p_l \rightarrow p_m$ *where* $i < j < k < l < m$

i.e., the processes are to be given in the order and in the correct proportions.

Feature Absolute Measurement (FAM)

The feature f_i may have many attributes and their values like represented for its measurement as the first dominant element, second and third dominant respectively and the degree of dominances as the first, second and third.

f_i *(first element, second, third, degree-first, degree-second, degree-third)* $= f_i (x, y, z, x', y', z')$

$$x' = x/\Pi x_i + \Pi y_i + \Pi z_i$$
$$y' = y/\Pi x_i + \Pi y_i + \Pi z_i$$
$$z' = z/\Pi x_i + \Pi y_i + \Pi z_i$$

Feature Additive Assessment (FAS)

Once the other materials or minerals are doped and added to the core material, the assessment of these combinations can be done with the following matrices multiplication for example. The purity level of the predicted chemical and the timing of the doping or additional elements determine the order or size of the matrix. The maximum and minimum levels of addition or doping required achieving the desired chemical characteristics of the fused features.

$$\begin{bmatrix} bestexpected & worst & default \\ catalystamount & reactant & timing \\ typical & negative & repeation \end{bmatrix} \bullet$$

$$\begin{bmatrix} variance_{purity} & accuracy_{dop} operat.temp \\ maxi_{permitted} & mini_{needed} mini \\ mini_{allowed} & maxi_{permissible} maxi \end{bmatrix} >= 0 \text{ for all } a_i * b_j$$

The Shapley value of a feature is its contribution to the compound or coalition, weighted and summed over all possible feature value combinations.

Individual Shapley Values

$$\varphi_{i(\text{material feature value})} = \varphi_j(val) = \sum S \subseteq \{x_1, \dots x_p\} / \{x_j\} |S|!\ (p - |S| - 1)!\ p!$$
$$(val(S \cup \{x_j\}) - val(S))$$

Interaction Shapley values

$$\varphi_{i,j} = \sum_{C \subseteq \{i.j\}} |C|!\,(M - |C| - 2)! \frac{\delta_{i,j}(C)}{2(M-1)!}$$

The Feature Fusion Explanation (FFE)

The cheminformatics-based causality for the emergent features of the compound can be used for fusion purposes. The reaction level values, process, constituent values and the operating environments are determining the outcome emergent properties of the materials. Function h_m maps a coalition to a valid instance. For hierarchically added features, h_m maps to the feature values of m. For absent features (0), an h_m map to the values of a randomly sampled data instance with and without Catalysts are shown as below:

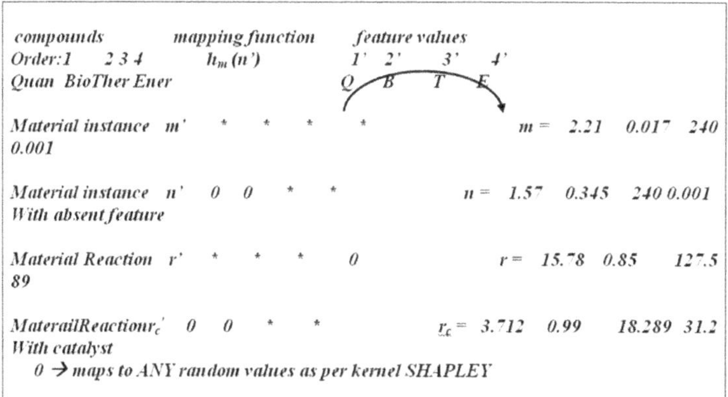

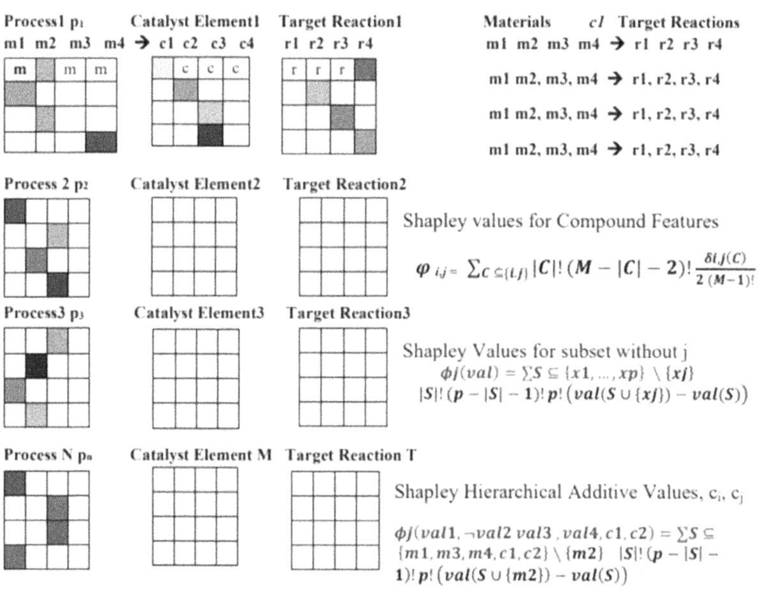

Lundberg had identified a kernel method for the Shapley compliant weighting techniques SHAP kernel can be rewritten for Idol which is a combination of features as,

$$Idol(c) = \frac{[M-1]}{(M|C'|)|c'|(M-|c'|)}$$

$$Idol(c) = \frac{[M1-1][M2-1]}{(M1M2|c1c2|)|c1c2|(M1M2-|c1c2|)}$$

Where M is the additive elements count or the size of the compound and $|c'|$ is the number of present features in instance c. Hence the $c1c2$ can also be considered with a different combination with and without the essential number of catalysts. The Idol Material Descriptive Language (IMDL) has been proposed in box text.

<General Data> place, fabricated-by, time, kingdom, size, weight, height, made-up-of
< Material Data> minerals, properties, foundry techniques
< Composition Data> sample composition and side effects
Historical Data – Source, literature or epigraphy
Damage Log – Rejuvenate and corrective repairs
Structural Data – Weight, volume, shape, height, specialty
Logistics Data – Carried by more than five people
<Beliefs> belief 1, name, why, what, where, when
belief 2, king, diety, power, curse, gift
belief 3, usage, benefit, actions, reactions, curation, medicinal
<Chemistry> Mineral set, Process set, Curing set, Maintenance tips,
<Documents> Literature, Epigraphical notes, Brass plates, Palm leafs
… … … … … … … …
Utility Data – Temple daily usage

Categorization of Materials based on Cheminformatics for Idols as Case Studies

The believer's beliefs based on the different materials, their compositions and their effects on the physical and psychological emotions of the human biological systems can be categorized as separate belief categories. Social or cultural beliefs are that the worship of idols made up of these five metal compositions will bring an improvement in social status, money inflow, health improvement, self-confidence and peace of mind to the people. These materials are primarily gold, silver, zinc, tin, iron, lead, rhodium and copper. The proportionate compositions of these materials with various chemical attributes and their thresholds play very important roles. The decision-making processes are based on these attributes and their exposure of these composites to other environmental factors. Chemical categories are created by separating these components. Because of its biochemical and electromagnetic properties, such as high conductivity, high melting points with great physical and chemical resistance, and optical qualities, gold and silver metals are employed in modern life. Silver is still used in X-rays and other medical procedures. Even before the discovery and capacity to make contemporary antibiotics, our ancestors employed silver for its natural antibacterial properties. Due to the increasing rise of antibiotic-resistant germ strains, researchers are once again focusing on silver as a critical component of future defenses to safeguard human health [2]. The healthcare industry has been utilizing platinum and ruthenium compounds to make pacemakers and defibrillators and to treat cancers. The documentary category, further on in this chapter, explores historical incidents, literary evidence and possible outcomes documented for the materials in those idols for reasoning purposes. These documents are engraved as metal carvings, written on palm leaves and epigraphically evidence written in ancient languages. Three different types, B matrix, C matrix and D matrix are formed in a qualitative manner using categorical theories and mapping between these categories. From the computational perspective, category theory focuses on a mathematical structure with nodes and arrows where the nodes are objects and the directed arrows are morphisms. These types of structures will have two properties, associativity of morphisms and identity for each object [17]. This data or information is called *categorical data* as compared to conventional numerical data. Many have developed many new classifications of material substances with reduced context, and discipline-specificity [18]. The curative properties of radioactivity have been practiced, and certain elements are absorbed differently by different organs, such as iodine, therefore iodine-131 is used to diagnose and treat thyroid cancer [19].

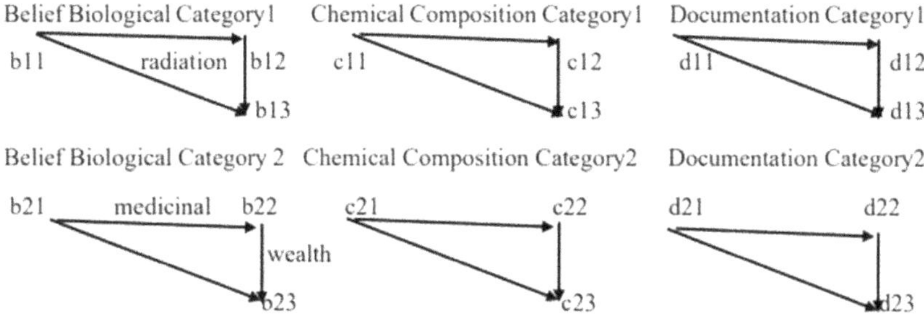

Figure 5: Belief chemical documents categorical data mapping diagrams

There are several current classifications of material substances, but no comprehensive classification that is adequate for any given application was found, as in the linked publications. Furthermore, they focus on material constituents that are primarily employed in engineering contexts, and they appear to be governed by their combinations and processes, with the eventual outcome they anticipated in [18]. The mapping, composition, and opposition of morphisms can be seen in Fig. 5. Consideration of opposite categories leads to a principle of duality and in the proposed domain, the relations among the objects of beliefs, chemical compositions and the documents can be represented as below. Within each category, a single mapping f11 without an apostrophe and across the categories, the f11' with an apostrophe are used to represent the various levels of mapping. These mappings are categorical data in a qualitative style or information content.

f11' g11'
Belief B11 $\rightarrow$ Chemical C11 $\rightarrow$ Document D11; composition of $g \circ f$

f11' g11'
Belief B11 $\leftarrow$ Chemical C11 $\leftarrow$ Document D11; opposition of $f \circ g$

The three categorical data can be mapped with the help of functors and objects within a category can be mapped using morphisms. The belief starts from the basic facts that the universe inclusive of all stars and the earth is virtually within the human functional body system and the human body is physically in the universe. The universal energy is constituted and contained within the five elements, soil, water, fire, air and space. Human energy is constituted by the minerals within the plants and the five metals covered under the soil. The basic belief is that with the various combinations of minerals from soil and plants, humans can control and command the entire universal entity. The metals within any human body system map and try to connect the minerals and metals within the outside world. This belief is extended by the belief that wearing the ornaments that are made up of five metals or "iympon" in Tamil or "panchalokhas" in the Sanskrit language will activate a strong binding and connectivity with the external energetic natural and cosmic resources. As a gift, utensils and containers made up of copper, brass, and bronze have been given to any new family to lead their life with these metals and presented decorative items and kitchen vessels made up of silver, gold and tin as a practice followed from ancient times. In ancient times, these metals were used as a base ornament mounted in different jewels with more precious stones and were offered to the Almighty during prayers. They believed that these extra-terrestrial forces like electromagnetic forces, cosmic forces and gravitational forces would be refracted through these minerals to the people who worship in these shrines and temples.

Chemical Composition Category

To fabricate and cast these impon idols, these idols will be made up of 85% copper, 13% brass and 2% graphite or carbon. Only a small amount of gold and silver will be added or doped into the composition to make it as a combination of five metal idols. The amount or weight of silver and gold will not be counted for the total overall weight of the sculpture. By wearing ornaments or jewels like ear rings or bangles that are made up of these metals, gold, silver, copper, zinc, iron or tin, the human body will be an attractive and metallic pulling force through these combinations of minerals. Metal combinations made up of these five minerals in different planned compositions act as a catalyst for all physical activities through appropriate energies in a biological human system. The bodies that are wearing these five metal ornaments have induced memory power, creativity, attractive attitude, wisdom, smartness, and awareness in the human biological body. The top of the temple tower is fitted with 3, 5, 7, 11 and 13 pots made up of impon that are called "kalasam" for worship. Pouring pure river water with all essential flowers and cosmetic minerals, the water will then be shared among all the people, to get some mineral power. Figures 6a, b, and c are collected through official documents of the Ministry of Hindu Temples to demonstrate the importance of categorization of metals as Impon.

Figure 6a: Sirkazhu, Kondal temple idols

Figure 6b: Four Nayanmars

Figure 6c.: Amman idol

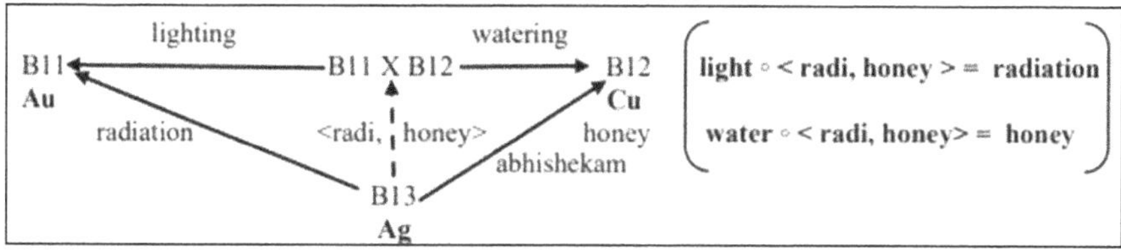

A "morphism of categories" is a functor. Across categories, there are functors in the feature fusion processes where F is the functor and identities of the objects or materials or features.

$F(g \circ f) = Fg \circ Ff$, $F\,idA = idF\,A$. The functor brings the causal reasoning in the chemical compositions and in the available literary documents.

Causal

Material Functors_{causal} → Causal mapping across categories B,C,D.

The categorical theory is concerned with a method of changing one functor into another while keeping the internal structure and morphism composition of the categories in question.

Evidential

Material $\text{Transformations}_{evidence}$ → Evidences epigraphical and Literary B, C, D

If epigraphy E and literature L are two functors across Chemical and Document categories, then the group of material transformations can be represented as

$$\eta \text{ from E to L such that}$$
$$\eta \text{ MATERIAL1} : \text{E (MATERIAL1)} \rightarrow \text{L (MATERIAL1)}$$

The commutation diagram of categorical data in the computational material science diagram based on the material features and evidences is shown below:

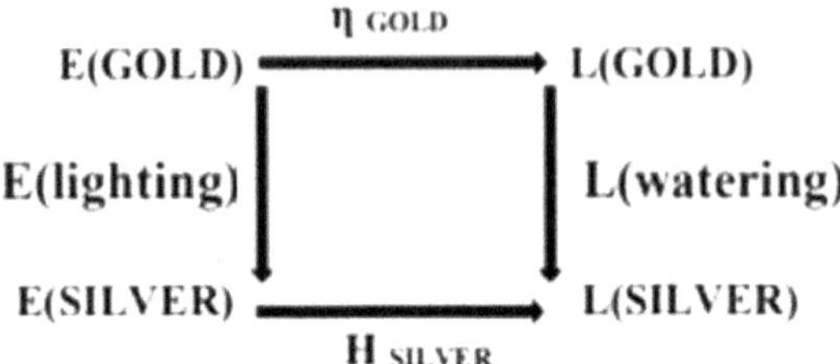

These levels of many such commutations are possible for different combinations of materials for the selection and fusion processes to reach the final explanations of the compounds. With natural transformations across functors, it is possible to determine the unique material and its additive materials and the hierarchical processes on which the features of the end target compound can be found out. These categorical data points can be used for supervised learning with the help of their respective Shapley values.

In Nagapattinam district of Tamil Nadu, a South Indian State that has more than 10000 temples with ancient idols and impon statues for all divine entities including disciples. Near the famous city, Sirkazhi, a small village called Kondal, there is a very old Murugan Temple that is approximately 800 years old approximately where, impon statues for all gods and disciples can be found as shown in the Fig. 1. In Ariyalor district, from Suthsmalli Varadharajaperumalkoil of Tamil Nadu, more than 28 idols made up of Iympon have been recovered from USA by the crime branch of Government of India as shown in Fig. 4. The idol or "vigraha" (Sanskrit) is made up of nine minerals called navaphasanam (Sanskrit). The panchamirtham, milk, sandal, honey and water used for abhishekam, a ritual adopted are believed to absorb these minerals to a very smaller amount. These micro nano molecular amounts contain curative properties. Another instance is the back of Tirupati Balaji Vigraha always remains moist, despite the priests labouring to keep it dry. The idol is resistant to the volatile chemical reactions of camphor, and bears no marks, even though it remains smeared with the substance most of the time [4]. In Southern India, the material used predominantly for "murthi" is black granite, while the material in North India is white marble. The black and white deities are changing the socio-cultural behaviour of those regional people very differently. With the synthesis of whole new families of organic and inorganic molecules containing spin-1/2 and spin-1 degrees of freedom, the new method known as "quantum magnetism" has seen a rebirth. The interplay between interactions and disorder in quantum systems within diverse materials is being studied using computational material science methods and paradigms, which will provide insight on phenomena. Figures 7a, b and c are collected through official documents of Ministry of Hindu Temples to demonstrate the importance of categorization of metals as Navapasanam.

Figure 7a: Kumbakonam Amman Idol

Figure 7b: Madurai District Chithira Vallaba Perumal Koil idols

Figure 7c: Uthirakosamangai Kuruvithurai Maragatha Nataraja Statue

Emerald is a gemstone and it is a sub variety of the mineral beryl ($Be_3Al_2(SiO_3)_6$) chemically and chromatically coloured green by trace amounts of chromium and/or sometimes vanadium. Figures 8a, b and c are collected through official documents of the Ministry of Hindu Temples to demonstrate the importance of categorization of single mineral as a rare/special category.

Figure 8a: Palani Murugan

Figure 8b: Thiruvidai Maruthoor, Kumbokonam

Figure 8c: Navapashana idol at Palani

This is made by a Siddhar called Bohar considering nine planets, 27 stars, electromagnetic radiations and multi crore minerals from plants juice mixture. The nine paashanam or poisons are Sathilinga Rasam, Manosilai, Arithaaram, Veeram, Sulphur, Pooram, White paashanam, Kauri Paashanam, Thotti Paashanam.

Documentary Evidence Category

In this category the Siddhar poems and palm leaves, the documentary evidences for the composition of nine different poisonous minerals or juices are added to form the "navapaashana". They are (Gowri paashanam) arsenic penta sulphite, (Kandhakap paashanam) sulphur, (Seelaip paashanam) arsenic disulfide, (Veerap paashanam) mercuric chloride, (Katchaalap paashanam) no proper official information is available, (Vellaip paashanam) arsenic trioxide and (Thottip paashanam) no proper official information is available, (Soothap paashanam) mercury, (Sanghup paashanam) no proper official information is available. It is a practice to plant a stone on the sixteenth day of the death of a person which is also called Paashanam. Eighty-one Siddhars including Bogar, the lead mixed eighty-one pashanams and converted into nine chemical composition mixtures. There were 18 Siddhars who were on the top list and there were 81 disciples who burnt the navapaasham by using nine types of fuels and filtered nine times ($9 \times 9 = 81$) to comply with the number nine at all levels. Since ancient times Indians have had their own rich knowledge of making and using of ornaments and utensils made of silver, gold, bronze, iron, and copper. Metals play an unimaginable role in bridging the human body with the cosmos. The ancestors used metal ornaments where gold and silver are used to cure disease. Some metals and minerals are used to cope with weaknesses; and so gold, silver, pearls, etc. are used by Ayurvedic doctors as medicines. Further, it is believed that by using Panchaloha (mix of five metals) one can get rid of evil things in life. Salt, camphor and soil are used in different ways to cure disease and cover up weakness and illness [21]. Gold and silver play an important role in human life. People who wear gold and silver are protected from problems of heat and infections. Wearing ornaments made of metal also indicates one's social status, so Indians either rich or poor wear gold and silver in more or less amounts. Married women wear gold mangalyam for their good health. Babies are embellished with silver anklets and bracelets for good health. Ancient Indians made copper and bronze statues of gods and goddesses. There are two types of evidence for the antiquity of mining and metallurgy in India: (a) literary evidence and (b) archaeological or archaeo metallurgical investigations in mining sites, equipment, slags, and other artefacts, as well as dating the artefact [22]. Wearing jewellery made of this alloy is thought to provide life balance, self-confidence, good health, wealth, prosperity, and peace of mind. Brass, gold, copper, silver, and zinc combine to make Panchaloha. The five also figuratively represent the five components that sanctify the cosmic centre of religious imagery. This alloy has provided a firm basis for what is commonly referred to as a bronze statue when melted in the proper proportions to generate a dazzling metal (iconic sculpture). The metals used are 4 parts of silver, 1 section of gold, 8 sections of brass, 8 sections of copper, and a small amount of iron.

This proportion is simply intended to give a general idea and may change depending on location. The material culture in rural India in general is far richer scientifically, and a logical explanation of all these is possible. However, modern educated people fail to understand this, while the uneducated or scarcely educated rural part of India follow it blindly without trying to understand the scientific reasons behind these cultures and customs of using materials. We people should be conscious about things like Pancha Klesha going inside our body and mind and also about the things outside the body such as plants, trees, animals, five elements, panchaloga. Each and everything in nature has a reason, and connection among each other, everything is interrelated. Therefore, it is very important to perform detailed research to explore the scientific purpose of those materials, to find the reason in what way they heal things and what is the actual science behind it. For categories C (chemicals), D (documents) define the product category C × D as follows. An object in C × D is a pair of objects from C and D, and an arrow in C × D is a pair of arrows from C and D. Identities and arrow composition are defined component-wise that can be represented as

$id(A,B) := (idA, idB),$

$(f, g) \circ (f', g') := (f \circ f', g \circ g')$ *as in material science domain, these may be written as,*

$feature_{(Gold, Silver)} := (feature_{Gold}, feature_{Silver}),$

$(lighting, watering) \circ (heating, covering): = (lighting \circ heating, watering \circ covering)$

The objects have morphisms as functions between them and categories have morphisms as functors between them and the functors have natural transformations between them as shown in Fig. 9.

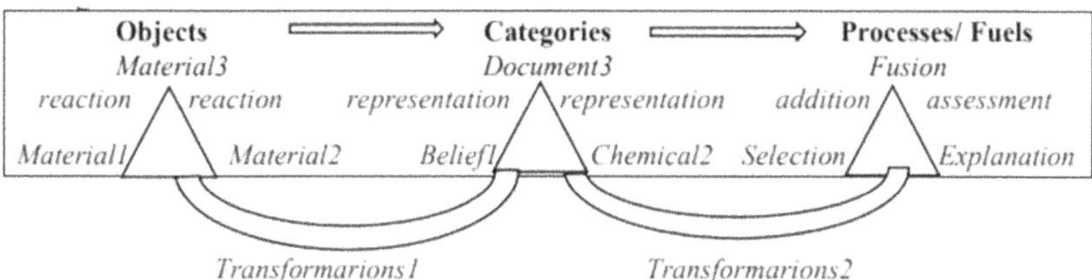

Figure 9: Function functors transformations in computational material science

Cheminformatics Based Logic

Deductive logic or inductive or intentional fusion-based logic can be derived using the Cheminformatics. This CBL logic can be enumerated using the aggregation and fusion of data from three sectors, History, Chemistry and Literature. Hardy, on the other hand, highlighted that alternative space-time metrics require a larger number of identifying points in order for the light cones to line up, at least locally. These locations, like space-time, form a reference frame in which causality appears to be definite. Once the base material is added with an additive in a sequence manner and if the processes are followed sequentially then one or more medicinal properties will emerge from the compound. Table 2 shows the various additive materials and process reasoning for their curative features.

base material |– additive minerals |– hierarchical process |– medicinal properties

$(bm_1, bm_2, bm_3, \dots bm_i), (am_1, am_2 \dots am_j), (hp_1, hp_2, \dots hp_k), (med_1, med_2, \dots med_p)$

The process sequences and the beliefs of these materials can be represented as process reduction calculus or process sequence calculus as shown below (p. 87) by a sample expression.

Table 2: Additive materials and process reasoning for their curative features

Serial number	Additive materials	Process names	Excrement Fuels	Hierarchical filters	Minerals reasoning
1	Avarai Chittamurutti	Heating	Kadai	Final stage	Carbon
2	Usil, Ilandhai	Washing	Kowdri	Preliminary	Potassium
3	Iluppai, Tamarind	Watering	Kukudam	First Stage	Magnesium
4	Neem, Pooarasu, Arasu	Grinding	Varadam	Second	Iron
5	Nuna, Vanni, Mavilangam	Heating	Peacock	Third	Zinc
6	Nelli, Vembu, Vila	Wet Grind	Elephant	Fourth	Copper
7	Usili,Vel,Konrai, Vengai	Dry Grind	Sand	Fifth	Bio Proteins
8	Panai, Thennai, Vembu	Mixing	Earth	First	Potassium
9	Vel Vengai	Adding	Gram	Last Stage	Zinc

$$\frac{\Gamma|-\Delta, b_{m1} \quad bm1, \Sigma|-\pi}{\Gamma, \Sigma|-\Delta, \pi} \equiv \frac{\text{gold, silver, belief, watering, honey}}{\text{gold, silve, zinc belief heat, light}}$$

$|-$ is a set a belief or lierary Γ, Σ are sequence of meterials and Δ, π are sequences of processes in the denominator whereas, the M is an instant material

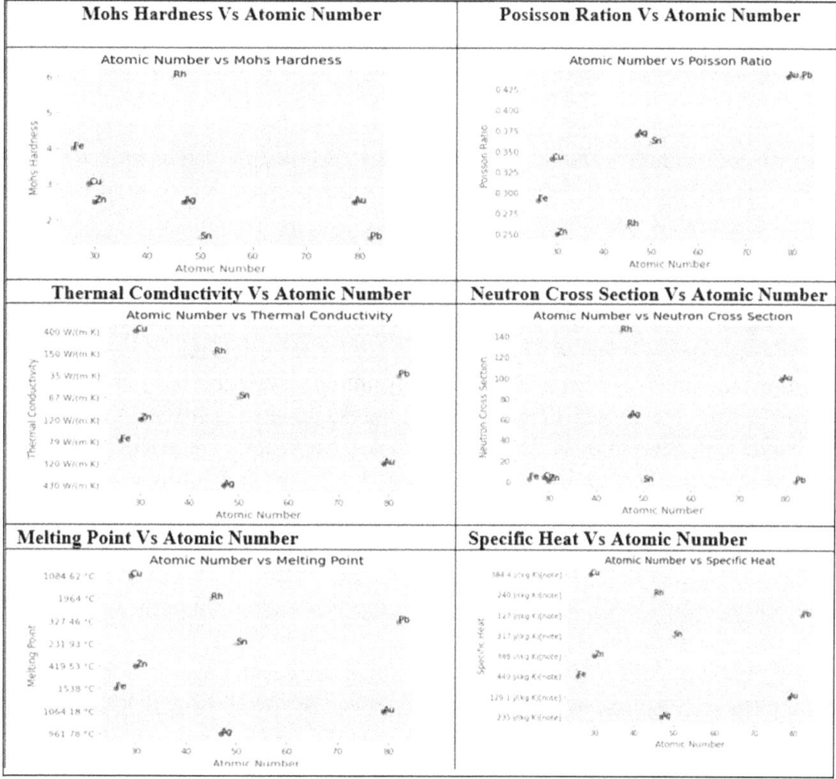

Figure 10a: Correlation among chemical different properties

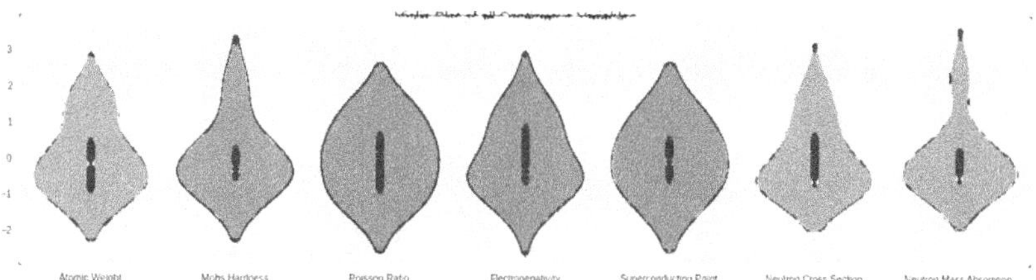

Figure 10b: Correlation among quantum different properties

The materials features are identified towards the target compound, the corresponding catalysts, combinatorics and process variables are maintained to maintain the biochemical quantum features of the mixture. The correlation among chemical and quantum properties are shown in Fig. 10a and 10b.

2. Conclusion

The computational material Science with cheminformatics feature descriptive language (CFDL) towards machine learning has been carried out in the research work. The inter-disciplinary research work focuses on the fundamental phases of material science like feature-based material discovery and material phenomena and their assessment in computational perspectives. Feature selection, assessment and explanation and the feature fusion processes are transformed into a computational problem based on the feature types and values. The feature-driven idol or temple sculptures are undertaken as case studies for the purpose of the preparation of digital documents for these precious assets. The actual preparation processes and procedures are not known clearly in the literary evidence; the study has to extend with more application of those minerals. The scope and the limitation of the research work are the non-availability of the process details through precious assets that were fabricated in ancient days with so much precision, the exact unknown mappings between the human physiological relations with the chemical properties of the minerals, as mentioned in the documents, is also a limiting factor in feature engineering of added explanations. The Shapley values of impon and navapaashaanam considering the processes, hierarchy and additive elements as per the proposed SHADE approach require high-performance computing due to the increased number of materials and processes. The most utilized materials called impon, like gold, silver, copper, iron, zinc and tin are considered for their chemical and quantum characteristics to be used for medicinal purposes based on a number of ancient literature as documents bearing evidence. The idol descriptive language is proposed to digitize the meta and micro specifications of these costly antiquities. A feature and reaction-based selection method has been proposed towards material selection and assessment. The samples are real from the existing complex sculptures available in India and this data was collected from the ancient literature and brass plates as epigraphical evidence. These three categories of categorical data were used to ascertain the novelty and originality of any idols or sculptures. The information or categorical data about the panchalikam or impon idols are taken as a case study and the biochemical radiation features are studied through data sheets.

Declaration of Interest

We wish to confirm that there are no known conflicts of interest among the authors in submitting the article to the Springer book edit and there has been no significant financial support for this work that could have influenced its outcome.

References

1. Wolchover, N. 2021. Quantum mischiefs rewrites the laws of cause and effect. *Quanta Magazine*, March 11.
2. Harmon, B.N. 2005. Computational materials science: Challenges, opportunities. *Journal of Physics*, 16(037): 273-276. Iowa State University, Ames Iowa, 50011, USA, Conference Series. Institute of Physics Publishing.
3. https://en.wikipedia.org/wiki/Spin-forbidden_reactions
4. https://indusscrolls.com/do-idols-emit-positive-energy/
5. Heine, T. 2014. Grand challenges in computational materials science: From description to prediction at all scales. *Frontiers in Materials*, doi:10.3389/fmats.2014.00007.
6. Himanen, L. et al. 2019. Data driven material science – Status, challenges and perspectives. *Advanced Science News*, 6: 1-23, 1900808.
7. Talirz, L., Kumbhar, S., Passaro, E. et al. 2020. Materials Cloud, a platform for open computational science. *Sci Data*, 7: 299.
8. Schmidt, J. et al. 2019. Recent advances and applications of machine learning in solid-state materials science. *NPJ Comput Mater*, 5: 83.
9. Ma, H., Govoni, M. and Galli, G. 2020. Quantum simulations of materials on near-term quantum computers. *NPJ Comput Mater*, 6: 85.
10. US Department of Energy, Office of Science. Quantum Materials—Harnessing the potential to revolutionize energy and information technologies.
11. American Chemical Society. 2021. Introduction: Quantum Materials. *Chem. Rev.*, 121: 2777-2779. https://dx.doi.org/10.1021/acs.chemrev.0c01322
12. Dane Morgan and Ryan Jacobs. 2020. Annual review of materials research opportunities and challenges for machine learning in materials science. *Machine Learning in Materials Science*, 71-103. www.annualreviews.org
13. Nargesian, F. et al. 2017. Learning feature engineering for classification. *Proceedings of the Twenty-Sixth International Joint Conference on Artificial Intelligence (IJCAI–17)*, 2529-2535. https://shap.readthedocs.io/en/latest/index.html#:~:text=SHAP%20.
14. Scott, M., Lundberg, Su–In Lee. 2017. A unified approach to interpreting model predictions, *31st Conference on Neural Information Processing Systems (NIPS 2017)*, Long Beach, CA, USA.
15. Utthara Gosa Mangai, Suranjana Samanta, Sukhendu Das and Pinaki Roy Chowdhury. 2010. A survey of decision fusion and feature fusion strategies for pattern classification. *IETE Technical Review*, 27: 293-307. https://en.wikipedia.org/wiki/Category_theory
16. Barbara, E. et al. 2018. Classification of material substances: Introducing a standards-based approach. *Materials and Design Journal*, 193(2020): 1-8.
17. https://www.epa.gov/radtown/radiation-used-nuclear-medicine.
18. Yezdani, U. et al. 2018. Nanotechnology in diagnosis and treatment of various diseases and its future advances in medicine. *World Journal of Pharmacy and Pharmaceutical Sciences*, 7(11): 1612-1633.
19. Waghmare, N.A. 2017. Historical importance of folklore material culture of Bidar. *Conference Proceedings of Social and Cultural Transformations through Ages*, 40-44.
20. Gandhi, S.M. 2014. Copper, zinc, lead ores – Their exploitation and metal extraction by the ancients in the north western India. *Journal of Geological Society of India*, 84: 253-266.

CHAPTER

6

Explicit Dynamic Crash Analysis of a Car Using a Metal, Composite Material and an Alloy

Pawandeep Dhall[1], Vijaykumar S. Jatti[2*], Nitin Solke[2]

[1] University of California, Berkeley, USA

[2] Symbiosis Institute of Technology, Pune, Maharashtra, India

1. Introduction

For a car to be structurally sound, its crash worthiness must be considered. As passenger safety, frame analysis, and material selection become increasingly important, it becomes an interesting field of study. In most cases, injury to an occupant is caused by collisions with other vehicles or objects on the road. No accident category causes more fatalities and severe injuries to the occupants of cars than frontal impact collisions. In 2015, car occupants accounted for 45% of deaths in Europe [8]. In [15], safety rules for vehicle design are discussed.

When a frontal impact occurs, the head, chest, and abdomen are the most critical areas for fatal injury. Additionally, most of the injuries resulting in disability occur in the legs and neck. Injuries and severity are strongly influenced by speed, car size, and several other parameters [1]. As a result of improvements in vehicle design, drivers can avoid accidents or even avoid accidents when they do occur; the vehicle should protect passers-by as well as the occupants against serious injuries.

Efforts are being made by governments and industries to produce better, greener, lighter cars while maintaining the safety of riders. Hydraulic shock absorbers are introduced in the front overhang to minimize deformation during the accident [2]. A chassis made of hybrid composite materials for light commercial vehicles is proposed in [3]. To improve the crash protection design of vehicles, researchers and industry are exploring trying out all the possibilities

*Corresponding author: vijaykumar.jatti@sitpune.edu.in

[4, 5, 6, 7]. Due to their strength and ductility, light composite materials along with aluminum, magnesium, plastic, or different variants of steel have been used in the construction of cars in recent years [13]. Crash tests based on numerical simulation are critical to validating the vehicle's deformation design strength [9, 10]. The most common types of crash tests are front-offset, front, side, and rollover [11, 12]. Advanced tools have been used for failure prediction for the crashworthiness of transportation vehicles. New failure laws were developed for different materials [14].

The costs of running physical tests are high, as are the times required to complete an experimental crash test, and the data used might be inaccurate. In this study, a frontal impact using an explicit dynamics code is simulated and analyzed to calculate the impact of the frontal collision on various barriers.

2. Methodology

This study was based on car crash tests of explicit dynamics case tests. The car material is modelled with three materials namely, stainless steel, composite and aluminum. Wall material is considered as nonlinear concrete. For each materials physical properties were defined.

The geometry is shown in Fig. 1. The car body was made of different materials and was 10 mm thick. The coarse mesh consisted of 10331 nodes and 9649 elements. Velocity was set at –30 m/s in x-direction. was added in the initial conditions of explicit dynamics.

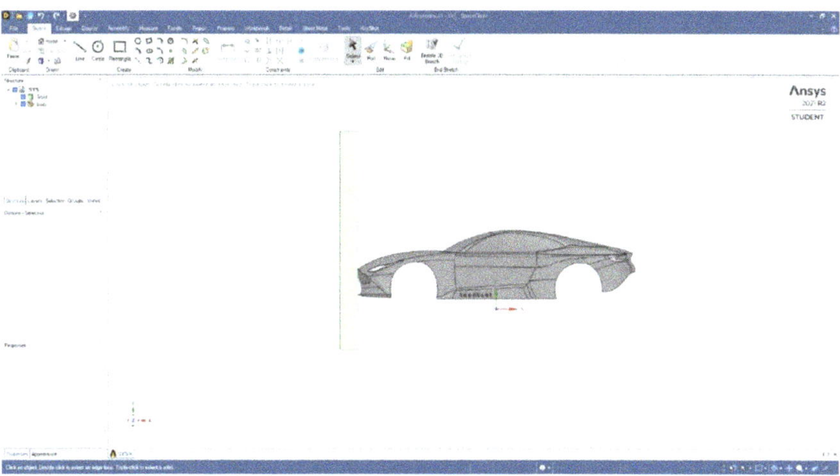

Figure 1: Geometry creation in space claim

Tables 1 and 2 depicts the material properties of composite material, aluminum, stainless steel, and concrete material, respectively.

Table 1: Material properties

	Composite	Aluminum	Stainless steel
Density kg/m^3	1857	2770	7750
Tensile yield strength Pa	4.401×10^8	2.8×10^8	2.07×10^8
Tensile ultimate strength Pa	4.401×10^8	3.1×10^8	5.86×10^8
Isotropic secant coefficient of thermal expansion/°C	1.688×10^{-5}	2.3×10^{-5}	1.7×10^{-5}
Isotropic thermal conductivity W/m °C	0.5523	237	15.1
Specific heat constant pressure J/kg °C	1069	875	480

Table 2: Concrete material properties

Density	2392 kg/m^3
Tensile yield strength	1.095×10^6 Pa
Tensile ultimate strength	1.196×10^6 Pa
Isotropic secant coefficient of thermal expansion	1.015×10^{-5} /°C
Isotropic thermal conductivity	2.933 W/m °C
Specific heat constant pressure	936.3 J/kg °C

3. Results and Discussions

The analysis was carried out using the student version of ANSYS, 2021 on a Windows Platform with Intel Core i7 with 8GB RAM. The following three cases are considered. The material assignment, dimensions of the bounding box and node-element statistics are shown below in Table 3. Same dimensions are chosen for different car materials. Thickness is chosen to be 1.e-002 m. The wall material is concrete with a mass of 13181 kg. The car materials chosen are

- Composite, epoxy/glass fibre, woven prepreg, biaxial 2 (264.11 kg)
- Aluminum (393.97 kg)
- Stainless steel (1102.3 kg)

Total deformation on the car body with composite material, aluminum and steel is shown in Fig. 2. It is observed that noticeable deformation occurs after 3 milliseconds, after which the crashing action begins.

Figures 6 and 7 illustrate how equivalent elastic strains and stresses differ among the three materials. The result summary is displayed in Table 4 in sequence of composite materials, aluminum and steel respectively. Figures 2, 3, and 4 shows the numerical results for all three materials.

The collision time and deformation are linearly related. The amount of deformation for the dynamic cars is higher, and then the deformation for the

Table 3: Nodes and elements properties

Bounding Box		
Object	**Car**	**Wall**
Length X	4.7923 m	0.3 m
Length Y	1.1576 m	3.6736 m
Length Z	2.0309 m	5 m
Statistics		
Nodes	10442	1620
Elements	9917	988

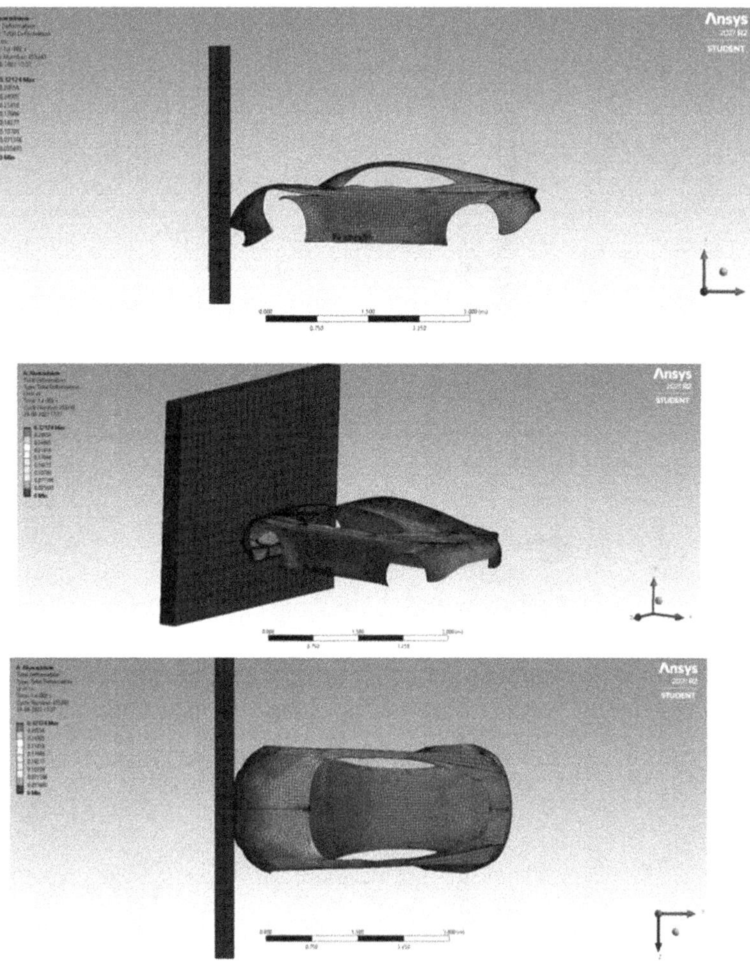

Figure 2: Total deformation composite material car body

Table 4: Result summary

Object name	Total deformation	Equivalent elastic strain	Equivalent stress
Minimum	0 m	0 m/m	8970. Pa
	0 m	3.0782×10^{-5} m/m	1.0891×10^{5} Pa
	0 m	0 m/m	32421 Pa
Maximum	0.32124 m	0 m/m	1.3075×10^{9}Pa
	0.36155 m	7.2068×10^{-2} m/m	3.7902×10^{9} Pa
	0.35023 m	0 m/m	8.4222×10^{9} Pa
Average	0.20821 m	0 m/m	4.3012×10^{7} Pa
	0.20467 m	2.5983×10^{-3} m/m	1.4919×10^{8} Pa
	0.20827 m	0. m/m	2.7463×10^{8} Pa

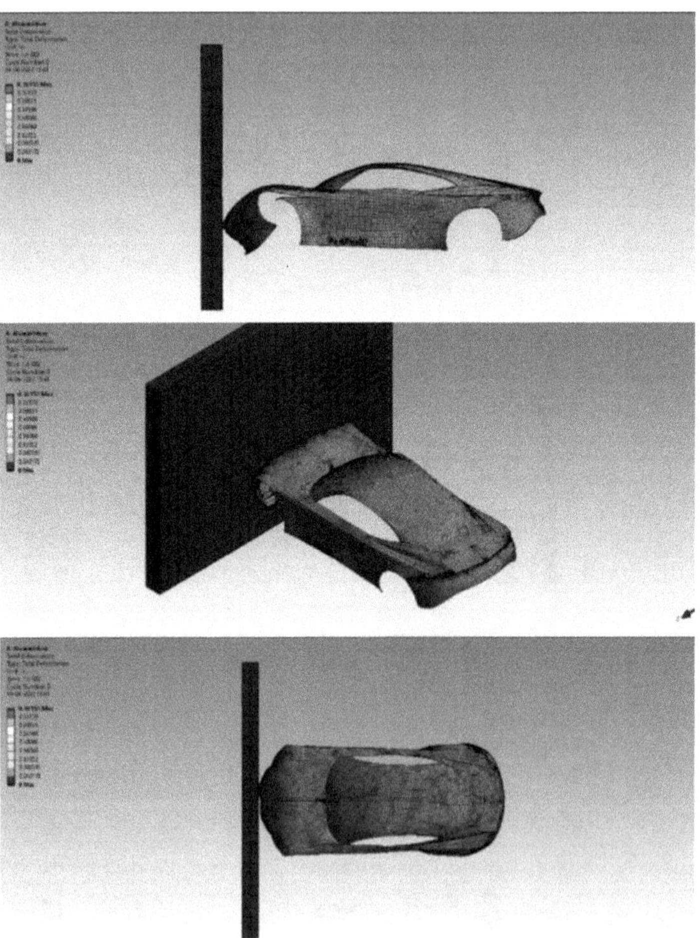

Figure 3: Total deformation aluminum material car body

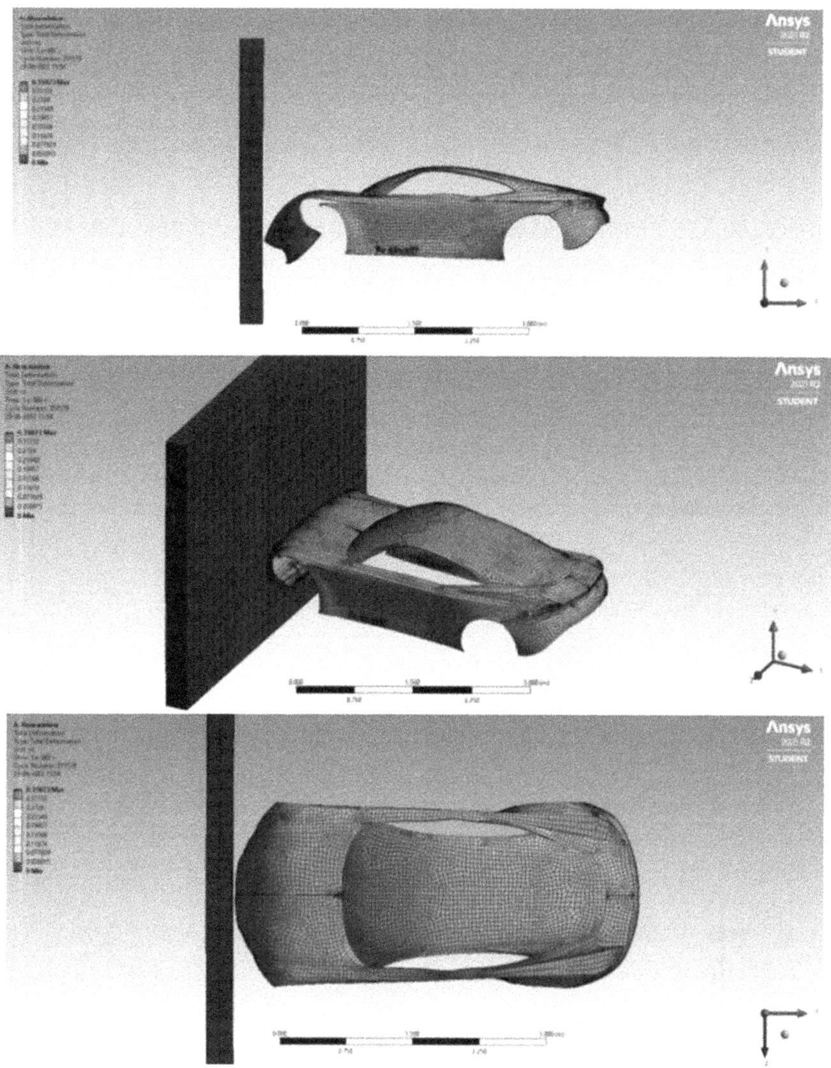

Figure 4: Total deformation stainless steel material car body

wall is next. It is, therefore, necessary to optimize the design obtained by using aluminum sheets since they are subject to high stresses and deformations. It is noted that composite cars are the lightest with 264 kg in mass whereas stainless steel cars are the heaviest with 1100 kg of mass. However, it is seen that stainless steel cars have the best readings to withstand crashes and composites have the most brittle nature. It is concluded that aluminum can prove to be a good material if strength is increased by performing some mechanical processes as it is lighter than stainless steel and stronger than composite material. Total deformation graph for different material of a car body is depicted in Figure 5, equivalent elastic strain graph shown in Figure 6, and equivalent stress graph is illustrated in Figure 7.

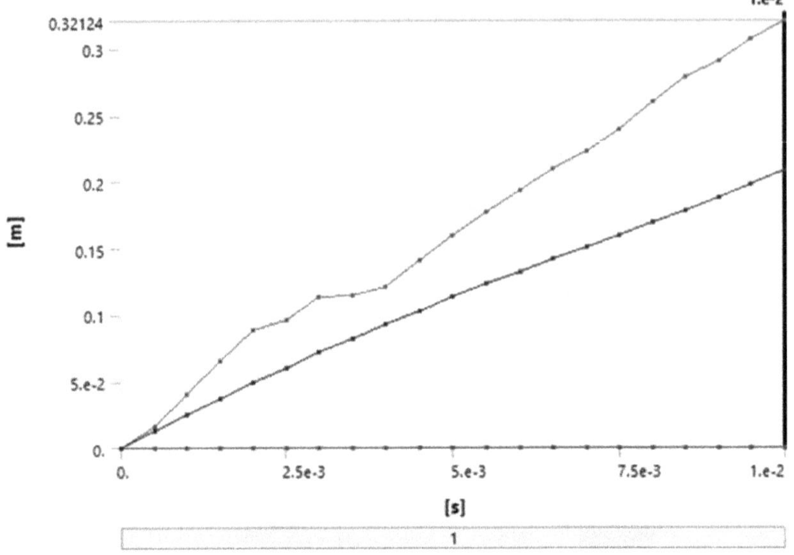

Figure 5: Total deformation graph for different material of a car body

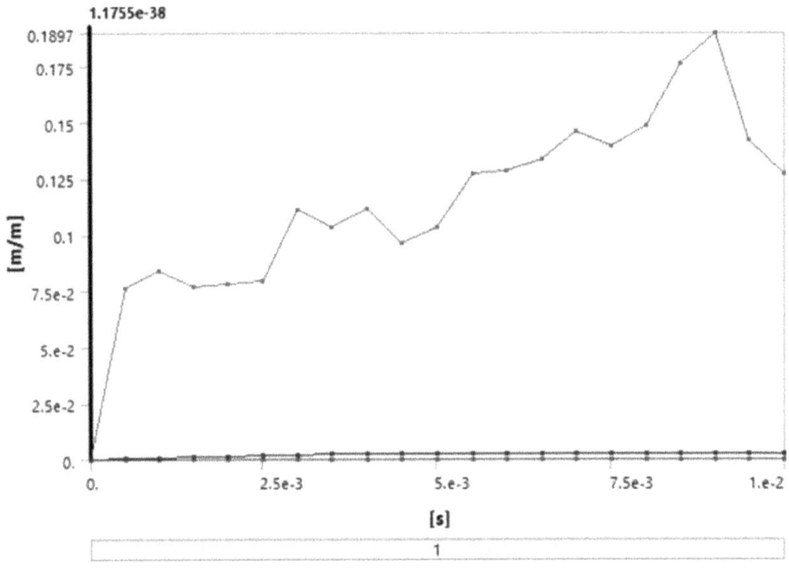

Figure 6: Equivalent elastic strain graph

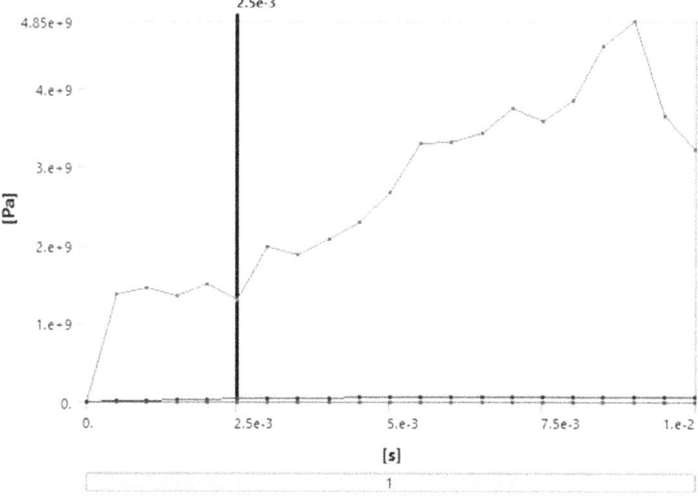

Figure 7: Equivalent stress graph

4. Conclusion

With hybrid and pure electric vehicles, mileage and weight have become trade-offs, so lightweight body parts are essential. Population growth and the rapid development of transportation infrastructures have heightened the demand for critical safety assessments of vehicles. In this paper, the ANSYS explicit dynamic method is applied to a simple car body structure to simulate a frontal collision at 30 m/s. Three different materials are considered in each case. In the study, it was found that moving cars caused more damage to their bodies than static walls. The pictures within this chapter illustrate how the vehicle deforms in each case. A future study is recommended for determining the best body structure and materials for car construction by modeling and analyzing the crashworthiness of various car body structures with hybrid aluminum material mixtures.

Acknowledgement

The authors would like to thank the Mechanical Engineering department, Dr. D. Y. Patil Vidyapeeth, Pune for providing the facilities to carry out this analysis.

References

1. Muhammad, A. and Shanono, I.H. 2019, December. Simulation of a car crash using ANSYS. *In: 2019 15th International Conference on Electronics, Computer and Computation (ICECCO)* (pp. 1-5). IEEE.

2. Balamurugan, R. and Sekar, D.M. 2017. Design of shock absorber for car front bumper. *International Journal of Science Technology & Engineering*, 3(9): 166-169.
3. Kiran, L., Kakkeri, S. and Deshpande, S. 2018. Proposal of hybrid composite material for light commercial vehicle chassis. *Materials Today: Proceedings*, 5(11): 24258-24267.
4. Pavan, B.S.N., Reddy, M.V. and Reddy, K.A. 2017. Design and crash analysis of car body using FRP material. *International Journal of Research in Advance Engineering Technology*, 6(2): 368-381.
5. Kiran, C.S., Sruthi, J. and Balaji, S.C. 2017. Design and crash analysis of a passenger car using ANSYS workbench. *CVR Journal of Science and Technology*, 13: 96-100.
6. Vulavapoodi Narayana, L. and Ramesh, K.V. 2016. Design and analysis of car bonnet by using PRO-E Abs ANSYS with FRP. *International Journal & Magazine of Engineering, Technology, Management and Research*, 3: 811-821.
7. Bhaskar, N. and Rayudu, P. 2015. Design and analysis of a car bonnet. *International Journal of Current Engineering and Technology*, 5(5): 3105-3109.
8. World Health Organization. 2015. *Global Status Report on Road Safety 2015*. World Health Organization.
9. Lin, C.S., Chou, K.D. and Yu, C.C. 2014. Numerical simulation of vehicle crashes. *Applied Mechanics and Materials*, 590: 135-143. Trans Tech Publications Ltd.
10. Sadhasivam, C. and Jayalakshmi, S. 2014. Simulation of car body by development of static and dynamic analysis. *International Journal of Automobile Engineering Research and Development*, 4(3): 1-6.
11. Ambati, T., Srikanth, K.V.N.S. and Veeraraju, P. 2012. Simulation of vehicular frontal crash-test. *International Journal of Applied Research in Mechanical Engineering (IJARME)*, 2(1): 37-42.
12. Babu, T., Praveen, D. and Venkateswarao, M. 2012. Crash analysis of car chassis frame using finite element method. *International Journal of Engineering Research & Technology*, 1(8): 1-8.
13. Masoumi, A., Shojaeefard, M.H. and Najibi, A. 2011. Comparison of steel, aluminum and composite bonnet in terms of pedestrian head impact. *Safety Science*, 49(10): 1371-1380.
14. Pickett, A.K., Pyttel, T., Payen, F., Lauro, F., Petrinic, N., Werner, H. and Christlein, J. 2004. Failure prediction for advanced crashworthiness of transportation vehicles. *International Journal of Impact Engineering*, 30(7): 853-872.
15. Hobbs, C.A. 2001. *Priorities for Motor Vehicle Safety Design*. Brussels (Belgium): European Traffic Safety Council.

CHAPTER

7

Optimizing Friction Stir Spot Welded ABS Weld Strength Using JAYA and Cohort Intelligence Algorithm

Rithvik Nair[1], Shreya Joshi[1], Kaustubh Dwivedi[1], Mandar S. Sapre[1*], Ashwini V. Jatti[2]

[1] Symbiosis Institute of Technology, Symbiosis International (Deemed University), Pune, India

[2] D.Y. Patil Institute of Technology, Savitribai Phule University, Pune, India

1. Introduction

Friction stir spot welding (FSSW) has emerged as a promising alternative to conventional welding techniques due to its many advantages such as improved joint strength, reduced distortion, and enhanced weld quality. One of the key factors that determine the success of FSSW is the optimization of process parameters as they can have a significant impact on weld quality and joint strength [1]. These parameters include tool geometry, rotational speed, and welding force, among others [2]. The application of FSSW to thermoplastic materials such as acrylonitrile-butadiene-styrene (ABS) has shown promising results [3]. ABS is a commonly used material in various industries due to its excellent mechanical properties, thermal stability, and resistance to impact and chemicals [4, 5, 3]. However, ABS welding can be challenging due to its high melting point and tendency to deform during the welding process. Therefore, the significance of FSSW parameters becomes even more crucial when ABS welding is involved. The optimization of process conditions, such as tool design, rotational speed, and welding force, can significantly impact weld quality, joint strength, and material deformation during the welding process [6]. In this chapter, we aim to investigate the significance of FSSW parameters during the ABS welding process and optimize the process conditions to achieve the most optimal possible weld quality.

*Corresponding author: mandar.sapre@sitpune.edu.in

Various studies have utilized statistical analysis, such as analysis of variance (ANOVA), to identify the most significant parameters that affect the mechanical properties of weld joints. However, due to the varying sets of parameters and operation ranges investigated across different studies, there can be some variation in the results obtained. A parameter that is found to be significant in one set of parameters and operation ranges may be insignificant in another.

In addition to studies aimed at optimizing mechanical properties [7], statistical models can also be employed for assessing and monitoring weld quality. In addition, they can also be employed for predicting and detecting weld defects at an early stage.

According to Adibeig et al. [8], rotation and traverse speeds have the most significant effect on the tensile strength of PMMA joints. Other crucial factors include traverse speed, rotational speed and pin diameter combination, axial force, tool diameter, and rotational speed. For instance, studies by Saeedy and Givi [9] on MDPE, Moochani et al. [10] on polypropylene, and Azarsa and Mostafapour [11] on HDPE sheets have reported that tilt angle, rotational speed, and tool temperature significantly affect joint strength. Moreover, the investigation conducted by Mosavvar et al. [12] revealed that rotational speed, tool offset (tilt angle in FSW of sheets), and traverse speed are the key parameters that significantly impact the yield strength in butt-joint FSW of HDPE pipes. A study by Bilici et al. [13] identified dwell time as the most critical factor for joint strength in HDPE, with rotational speed and plunge depth also playing significant roles. Similarly, in another study, Bilici [14] also conducted a study on polypropylene and concluded that dwell time has the greatest impact on joint strength, followed by plunge depth and rotational speed. Lambiase et al. [15] found that dwell time had the most impact on joint strength in their study of polycarbonate. They also observed that plunge rate, waiting time, and a combination of plunge rate and dwell time had significant effects on joint strength. In a separate study, Bilici et al. [16] found that the rotational speed influenced joint strength in HDPE the most, followed by the heat transfer coefficient of the tool material, dwell time, and plunge depth.

The paper is organized in a manner to present the results of the experiment on the FSSW of ABS polymer using a cylindrically shaped pin on a milling machine. The methodology section describes the experimental setup, materials, and process parameters utilized in the study. The welding process parameters, including spindle speed, plunge depth, and dwell time, were varied according to a predetermined experimental plan, which consisted of a series of tests. Tensile data collected during the test was then used to develop a mathematical model that could predict the response of the process parameters. Subsequently, the application of five different optimization algorithms, namely Genetic Algorithm (GA), Particle Swarm Optimization (PSO), Simulated Annealing (SA), CI, and Jaya Algorithm is discussed. The results section presents the UTS data for the friction-welded ABS polymer samples, followed by a discussion of the results and the insights gained from the study. Finally, the conclusion summarizes the findings of the study.

2. Methodology

In this study, experiments were conducted to investigate the significance of FSSW parameters while ABS welding. A cylindrical-shaped pin with a thickness of 3 mm was employed by the research group, as shown in Fig. 1a.

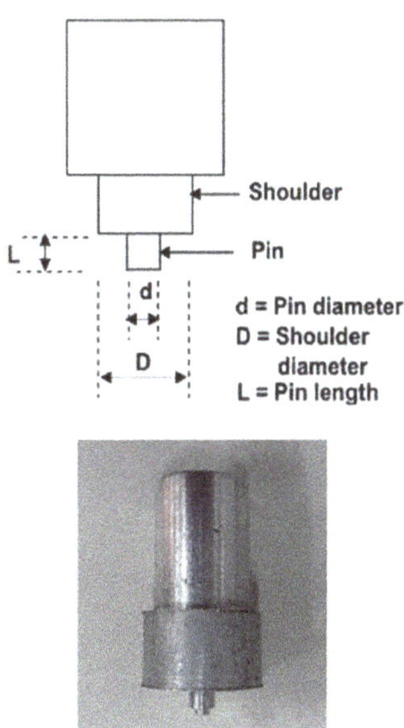

Figure 1a: FSSW Tool

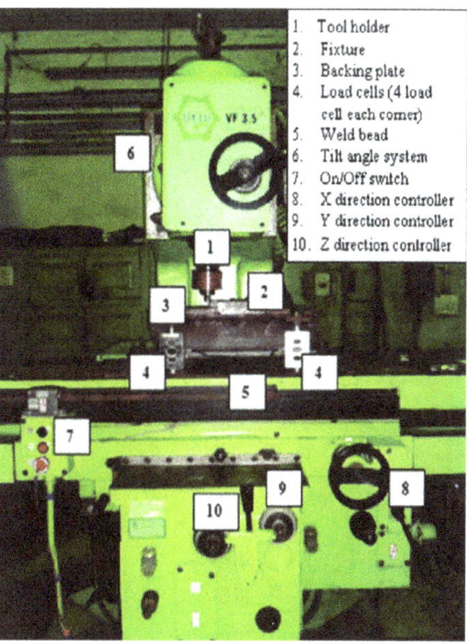

Figure 1b: Vertical milling machine

The simplicity and mechanical results of the cylindrical-shaped pin made it an ideal choice for the study. ABS, a thermoplastic material was used for the experiments because of its excellent mechanical properties, thermal stability, and chemical resistance. The dimensions of the ABS workpiece were 74 mm in width, 140 mm in length, and 3 mm in thickness. The vertical milling machine was used as the machine for the FSSW process as shown in Fig. 1b. The friction stir welding parameters for ABS were carefully selected and designed to ensure optimal welding conditions.

The effect of the parameters is discussed in Table 1. The first column represents the number of samples, the second column represents the spindle speed (rpm), the third column represents the plunge depth (mm), and the fourth column represents the dwell time (seconds). Spindle speed is one of the key parameters that affect the welding process as it influences the heat generated during the FSSW process. The plunge depth is the depth to which the rotating tool is plunged into the material, and it also affects the amount of heat generated during the welding process. The dwell time is the time for which the tool remains stationary at the bottom of the weld zone, and it influences the bonding strength of the

Table 1: Experimental layout and observed values

Sample no.	Spindle speed (rpm)	Plunge depth (mm)	Dwell time (secs)	UTS (MPa)
1	1000	0.8	40	1.528
2	1000	0.8	60	0.573
3	1000	0.8	80	0.961
4	1000	0.9	40	0.37
5	1000	0.9	60	1.12
6	1000	0.9	80	0.741
7	1000	1	40	5.419
8	1000	1	60	5.18
9	1000	1	80	3.009
10	1400	0.8	40	0.361
11	1400	0.8	60	0.688
12	1400	0.8	80	1.281
13	1400	0.9	40	2.113
14	1400	0.9	60	5.254
15	1400	0.9	80	2.313
16	1400	1	40	0.22
17	1400	1	60	0.326
18	1400	1	80	Weld not formed
19	2000	0.8	40	0.879
20	2000	0.8	60	0.153
21	2000	0.8	80	1.754
22	2000	0.9	40	Weld not formed
23	2000	0.9	60	0.364
24	2000	0.9	80	Weld not formed
25	2000	1	40	2.274
26	2000	1	60	Weld not formed
27	2000	1	80	0.67

weld. Overall, the experimental details in this study were carefully designed and executed to investigate the significance of FSSW parameters while ABS welding. The selection and optimization of these parameters are crucial for achieving the desired weld quality and joint performance, and the results of this study can potentially contribute to the development of more effective and efficient welding techniques for thermoplastic materials, particularly ABS.

The experiment involved performing double pass FSSW on 2 mm thick and 70 mm by 50 mm pieces of Acrylonitrile Butadiene Styrene (ABS) to form Butt Welds. The study aimed to understand the effect of process parameters, including spindle speed, plunge depth, and dwell time, on the welding process. This was achieved through three sets of experiments using a full factorial design of three parameters varying on three levels across 27 trial conditions.

Figure 2 depicts the workflow for the current study. After welding, laser-cut samples were sent for UTS testing. Figure 3a shows laser-cut tensile samples after the FSSW weld and Fig. 3b shows tensile samples after the tensile test. Nonlinear regression analysis was then conducted, based on which a non-linear model was developed. In this study metaheuristic algorithms viz. GA, SA, PSO, JAYA, and CI were used to optimize UTS. Conformational experiments were conducted based on the optimized value obtained from the tests to validate the model. This study aims to provide insight into the relationship between process parameters and mechanical properties of ABS welds, which can be used to optimize FSSW processes.

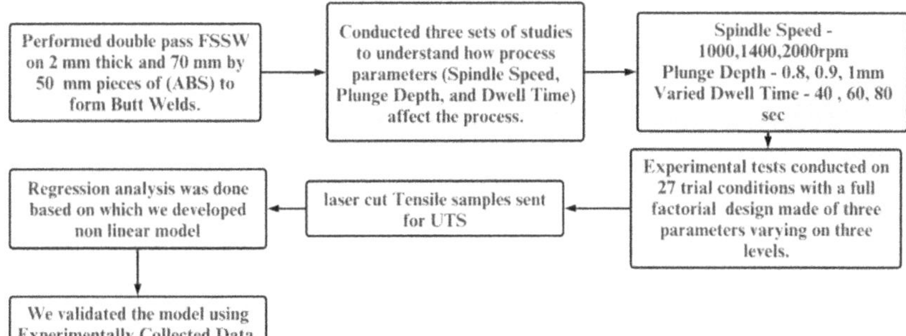

Figure 2: Workflow

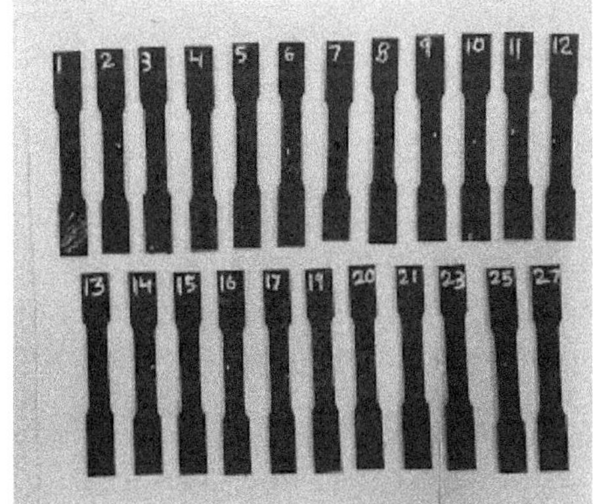

Figure 3a: Tensile sample before test

Figure 3b: Sample after test

2.1 Formulation of Equation

The first step was the formulation of a non-linear regression equation for UTS based on experimental values. The equation considers the desired outputs—UTS and inputs—spindle speed, dwell time, and plunge depth parameters. The equation was based on the physics of the process, as well as empirical data gathered from experiments. By using this equation and optimization approach, it was possible to significantly improve the performance of the algorithm and achieve more accurate results in the present study. Equation 1 depicts the nonlinear equation, where the coefficients associated with each variable and interaction term indicate the impact of each input on the output. The equation was used to model the relationship between the input variables and the output response and was utilized in the analysis and optimization of experimental results.

$$\begin{aligned} UTS = (&0.2088x_1 + 25.0406x_2 + 0.0119x_3 - 0.0004x_1' + 15.1412x_2' \\ &+ 0.000001x_3' - 0.2454x_1x_2 + 0.000035x_1x_3 - 0.0198x_2x_3 - 27.1536) \end{aligned} \tag{1}$$

where, x_1 = Spindle speed, x_2 = Plunge depth, x_3 = Dwell time

2.2 Optimization Algorithms

In this study, five optimization algorithms are used to optimize the FSSW process for ABS parts. The goal of optimization is to find the values of the parameters that would result in the best output, which is determined by a specific metric. The study uses GAs, simulated annealing, particle swarm optimization, JAYA algorithm, and CI algorithm to find the optimal process parameters. Figure 4 illustrates the flow of the JAYA optimization process adopted in the present study. Algorithm-1 and Pseudocode-1 depict the implementation of JAYA optimization.

Algorithm-1: JAYA optimization

Set the population size N, the maximum number of iterations iter_max, the lower and upper bounds of the search space for each variable, and the optimization objective function f(x). Initialize a population of N individuals randomly within the search space. Evaluate the fitness function f(x) for each individual in the population. Repeat until the maximum number of iterations is reached or a termination criterion is met: Find the best and worst individuals in the population. Update each individual in the population as follows: For each variable, generate a new value by adding a random number between −1 and 1 times the difference between the best and worst individuals to the corresponding variable value of the current individual. If the new value is outside the bounds of the search space, clip it to the closest bound. Evaluate the fitness function f(x) for each individual in the population. Return the best individual found during the search.

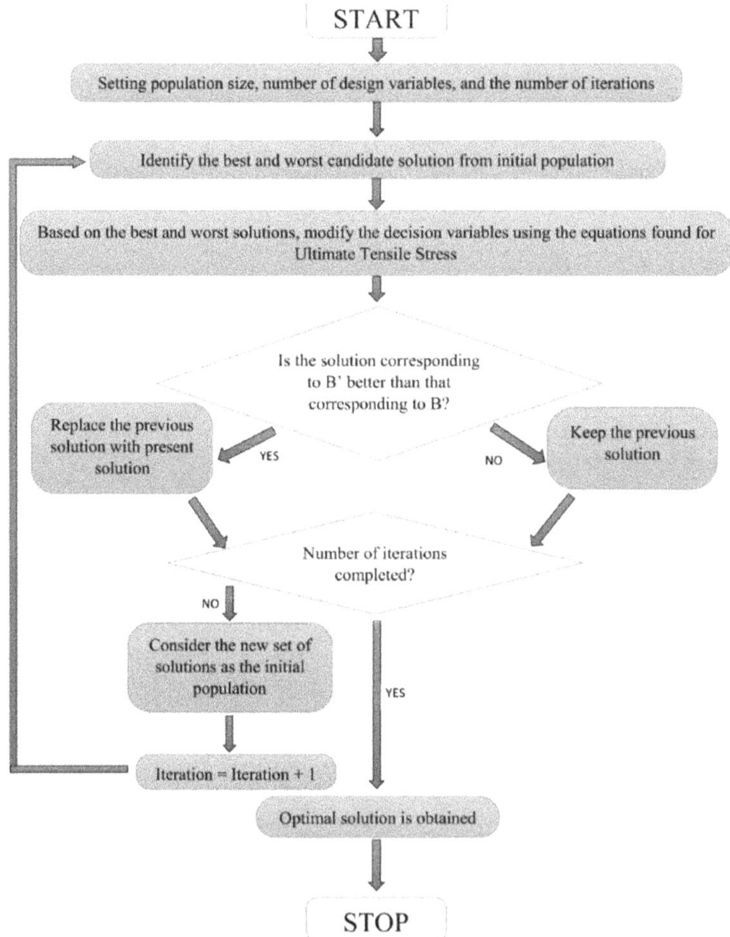

Figure 4: JAYA algorithm flowchart

Pseudocode-1: JAYA optimization

```
% Initialize population and evaluate fitness.
pos = lb + rand(N,D) .* (ub - lb);
fitness = evaluate_fitness(pos);
% Repeat until stopping criterion is met.
for iter = 1:itermax
   % Update worst solution using greedy strategy.
   [~, worst_idx] = max(fitness);
worst_sol = pos(worst_idx,:);
```

(Contd.)

```
worst_sol(worst_sol< lb) = lb(worst_sol< lb);
worst_sol(worst_sol>ub) = ub(worst_sol>ub);
pos(worst_idx,:) = worst_sol;
  % Evaluate fitness of updated population.
new_fitness = evaluate_fitness(pos);
  % Update best solution found so far and check stopping criterion.
  [best_fitness, best_idx] = min(new_fitness);
  if (iter>1) && (abs(best_fitness - prev_fitness) <tol), break; end
prev_fitness = best_fitness;
  % Display best fitness and plot convergence.
disp(['Iteration ' num2str(iter)': Best fitness = 'num2str(best_fitness)' , Best solution = [' num2str(pos(best_idx,:))']']);
plot(iter, best_fitness, 'bo');
  % Update fitness array.
  fitness = new_fitness;
end
% Return best solution found.
best_sol = pos(best_idx,:);
```

Figure 5 illustrates the flow of the GA optimization process adopted in the present study. Algorithm-2 and Pseudocode-2 depicts the implementation of GA.

Algorithm-2: GA

Set the population size N, the maximum number of generations G_max, the mutation probability P_m, and the crossover probability P_c.
Initialize a population of N individuals randomly within the search space.
Evaluate the fitness function f(x) for each individual in the population.

Repeat until the maximum number of generations is reached or a termination criterion is met:
Select parents for reproduction using a selection method such as roulette wheel selection or tournament selection.
Generate offspring by applying genetic operators such as crossover and mutation to the parents. Use crossover with probability P_c and mutation with probability P_m.
Evaluate the fitness function f(x) for each offspring.
Select individuals to survive to the next generation using a survival method such as elitist selection or rank-based selection.
Replace the current population with the selected individuals to form the next generation.

Evaluate the fitness function f(x) for each individual in the population.

Return the best individual found during the search.

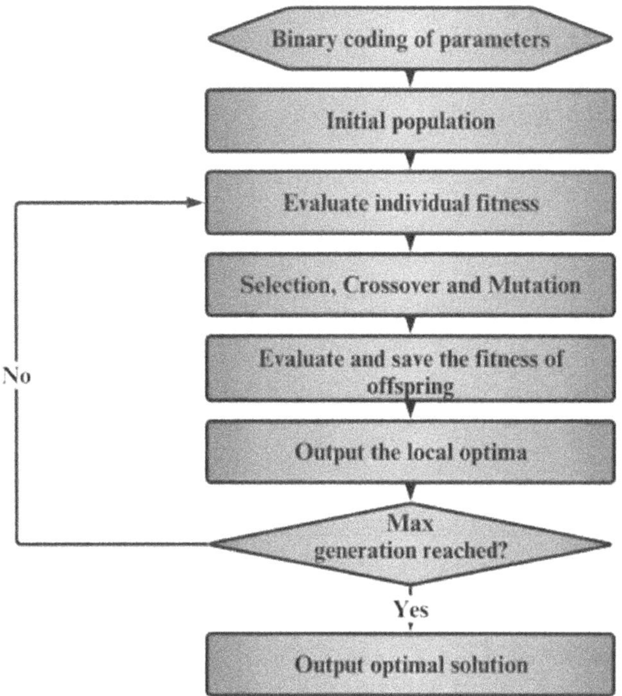

Figure 5: GA flowchart

Pseudocode-2: GA

```
% Initialize population of size N
N = 100;
population = rand(N,1);
% Calculate fitness of each individual in population
fitness = zeros(N,1);
for i = 1:N
   fitness(i) = calculateFitness(population(i));
end
% Repeat until termination condition is met
while ~terminationConditionMet()
   % Select mating pool of individuals based on fitness
matingPool = selectMatingPool(population, fitness);
   % Perform crossover operation with probability Pc on mating pool
   Pc = 0.8;
  for i = 1:2:length(matingPool)
     if rand() < Pc
```

(*Contd.*)

```
        offspring = crossover(matingPool(i), matingPool(i+1));
matingPool(i) = offspring(1);
matingPool(i+1) = offspring(2);
      end
    end
    % Perform mutation operation with probability Pm on mating pool
    Pm = 0.1;
    for i = 1:length(matingPool)
      if rand() < Pm
matingPool(i) = mutate(matingPool(i));
      end
    end
    % Replace parents in population with children
    population = matingPool;
    % Calculate fitness of each individual in population
    for i = 1:N
      fitness(i) = calculateFitness(population(i));
    end
end
% Return best individual from population
[~, ind] = max(fitness);
bestIndividual = population(ind);
```

Figure 6 illustrates the flow of the SA optimization process adopted in the present study. Algorithm-3 and Pseudocode-3 depicts the implementation of simulated annealing.

Algorithm-3: Simulated Annealing

Set the initial temperature T, the cooling schedule function S, the maximum number of iterations iter_max, and the initial solution x.
Set the current solution x_curr to the initial solution x.
Repeat until the maximum number of iterations is reached or a termination criterion is met:
Generate a candidate solution x_cand by applying a perturbation function to the current solution x_curr.
Calculate the objective function value f(x_cand) for the candidate solution.
Calculate the objective function value f(x_curr) for the current solution.
If f(x_cand) is better than f(x_curr), set x_curr to x_cand.
If f(x_cand) is worse than f(x_curr), accept x_cand with probability P = exp(-(f(x_cand) - f(x_curr)) / T).
Update the temperature T using the cooling schedule function S.
Return the best solution found during the search.

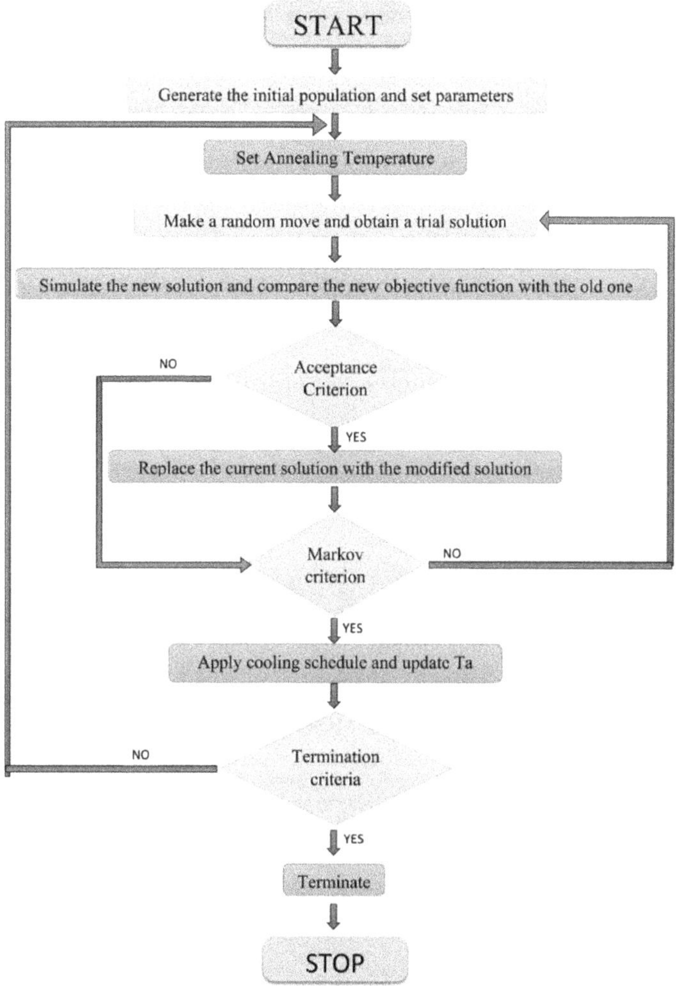

Figure 6: SA flowchart

Pseudocode-3: Simulated Annealing

```
% Set initial solution s0, temperature T, cooling schedule parameters alpha and
num_iter, and T_f
s0 = ...; T = ...; alpha = ...; num_iter = ...; T_f = ...;
% Set current solution s = s0 and best solution found so far as s
s = s0; best_s = s;
f_s = f(s); best_f = f_s;
% Repeat until T<T_f
while T >T_f
```

(Contd.)

```
% For each temperature, perform a number of iterations to explore the neighborhood of the current solution
for i = 1:num_iter
% Generate a candidate solution s' by applying a random perturbation to s
s_prime = perturb(s); f_s_prime = f(s_prime);
vbnet
Copy code
   % Accept candidate solution s' if it has lower energy
   delta = f_s_prime - f_s;
   if delta < 0 || rand() < exp(-delta/T)
      s = s_prime; f_s = f_s_prime;
      % Update best solution found so far
      if f_s<best_f
best_s = s; best_f = f_s;
      end
   end
end
% Cool down the temperature
T = T * alpha;
end
% Return the best solution found during the search
best_solution = best_s; best_obj = best_f;
```

Figure 7 illustrates the flow of the PSO process adopted in the present study. Algorithm-4 and Pseudocode-4 depicts the implementation of PSO.

Algorithm-4: Particle Swarm Optimization

```
Set the number of particles n, the maximum number of iterations iter_max, the inertia weight w, the cognitive factor c1, and the social factor c2.
Initialize the position x and velocity v of each particle randomly within the search space.
Evaluate the objective function f(x) for each particle.
Set the personal best position pbest of each particle to its current position x and the corresponding personal best value pbest_value to its current objective function value f(x).
Find the global best position gbest and the corresponding global best value gbest_value among all personal best positions pbest.
Repeat until the maximum number of iterations is reached or a termination criterion is met:
Update the velocity and position of each particle:
Update the velocity of the particle using the formula: v = w * v + c1 * rand() * (pbest - x) + c2 * rand() * (gbest - x)
Update the position of the particle using the formula: x = x + v
Evaluate the objective function f(x) for each particle.
```

(Contd.)

Update the personal best position pbest and the corresponding personal best value pbest_value for each particle if its objective function value is better than its previous personal best value.
Update the global best position gbest and the corresponding global best value gbest_value if any personal best value is better than the current global best value.
Return the global best position gbest and the corresponding global best value gbest_value.

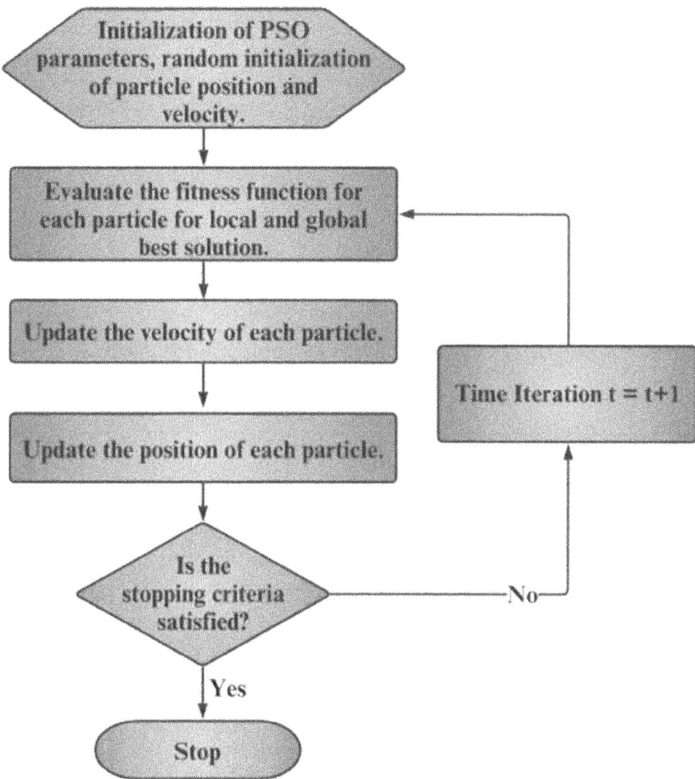

Figure 7: PSO flowchart

Pseudocode-4: Particle Swarm Optimization

```
function [gbest, fgbest] = PSO(fobj, lb, ub, n_particles, n_iterations)
   particles = lb + rand(n_particles,numel(lb)).*(ub-lb);
   velocities = zeros(n_particles,numel(lb));
pbest = particles;
fpbest = inf(n_particles,1);
```

(Contd.)

```
[fgbest, gbest_idx] = min(fpbest);
gbest = pbest(gbest_idx,:);
  w=0.7; c1=1.5; c2=1.5;
  for i=1:n_iterations
    velocities=w*velocities+c1*rand(n_particles,numel(lb)).*(pbest-particles)+c2*rand(n_particles,numel(lb)).*(repmat(gbest,n_particles,1)-particles);
      particles = particles + velocities;
      particles = max(min(particles,ub),lb);
      fitness = arrayfun(fobj,particles);
idx = fitness <fpbest;
pbest(idx,:) = particles(idx,:);
fpbest(idx) = fitness(idx);
      [fgbest_new, gbest_idx_new] = min(fpbest);
      if fgbest_new<fgbest
gbest = pbest(gbest_idx_new,:);
fgbest = fgbest_new;
      end
  end
end
```

The Pseudocode of the CI approach is presented in Pseudocode-5.

Pseudocode-5: Cohort Intelligence

Let
K Number of candidates ($k = 1, ..., K$)
S Sampling space
r Sampling space reduction factor
Initialize K, S, r
While
Generate the random solutions $\mathbf{X}$.
Evaluate function behavior $f(\mathbf{X}^k)$ for each candidate.

The probability p^k associated with every candidate k in the cohort is calculated as: $p^k = \dfrac{f^*(\mathbf{X}^k)}{\sum_{k=1}^{K}[f^*(\mathbf{X}^k)]}$

Using roulette wheel approach every candidate k selects behaviour to follow the behaviour with highest probability.
Every candidate k expand/shrinks the sampling interval $\boldsymbol{S}^c$ in its neighbourhood using sampling space reduction parameter r:

$$[\boldsymbol{S}^{k,lower}, \boldsymbol{S}^{k,upper}] \left[\mathbf{X}^k - \left\|\frac{\boldsymbol{S}^{upper} - \boldsymbol{S}^{lower}}{2}\right\| \times r, \mathbf{X}^k + \left\|\frac{\boldsymbol{S}^{upper} - \boldsymbol{S}^{lower}}{2}\right\| \times r\right]$$

Check for saturation

3. Results and Discussions

In FSSW, dwell time, spindle speed, and plunge depth are critical parameters that can significantly affect the quality of the weld. Several studies have investigated the effect of these parameters on the UTS of the weld. It has been found that an increase in dwell time and spindle speed can lead to an increase in UTS. On the other hand, an increase in plunge depth can increase both UTS. These findings suggest that the selection of appropriate values for these parameters is critical for achieving the desired mechanical properties in FSSW. These experimental parameters are consistent with those reported in previous studies investigating dwell time, spindle speed, and plunge depth effects on the mechanical properties of FSSW.

3.1 GA Results

This study utilized a hybrid optimizer, specifically a combination of GA and pattern search, to optimize the function. The GA component of the optimizer employed a crossover fraction of 0.9, which means that 90% of the genes on the parent chromosomes were used to create the offspring. The fitness scaling function used was the fitness scaling prop, which scales the fitness values based on their proportion to the overall population's fitness. The selection function was selection roulette, which selects individuals for reproduction based on their fitness values. The crossover function combines genetic material from two parent chromosomes with a heuristic approach. The crossover mutation used was mutation power, which adds a random value to the gene with a certain probability of introducing novel variations into the population. Using these settings, we obtained the ideal optimal fitness value of 5.4242 for the function UTS, as shown in Fig. 8. The hybrid optimizer allowed us to efficiently explore the solution space and find the

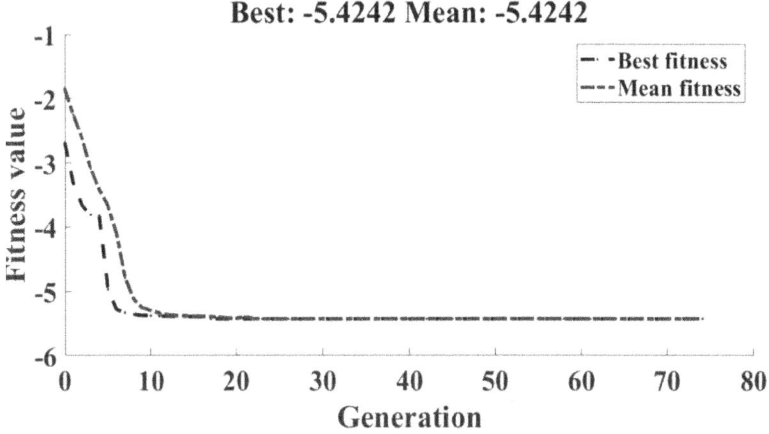

Figure 8: GA: fitness value versus generation plot

optimal values of the input parameters that resulted in the desired output. These results demonstrate the effectiveness of the GA pattern search hybrid optimizer in optimizing complex functions with multiple input parameters.

3.2 PSO Results

The PSO algorithm was configured with a minimum neighbour fraction of 0.25, which means that each particle has at least 25% of its neighbours in its local search space. The social adjective weight was set to 1.2, which controls the influence of the global best position on the particle's movement. The self-adjective weight was set to 1.49, which controls the particle's tendency to move towards its personal best position. Using these settings, we obtained the best fitness value of 5.4242 for our UTS, as shown in Fig. 9. The PSO algorithm allowed us to efficiently search the solution space and find the optimal values of the input parameters that resulted in the desired output. The results demonstrate the effectiveness of the PSO algorithm in optimizing complex functions with multiple input parameters.

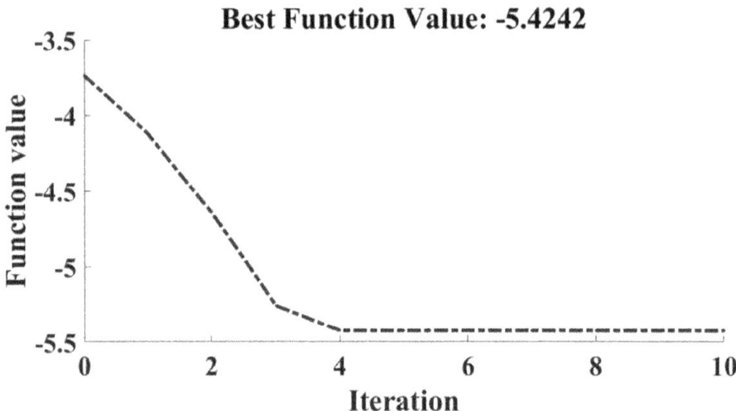

Figure 9: Particle swarm optimization: function value versus iteration plot

3.3 SA Results

We also used SA to optimize our function with multiple input parameters. The SA algorithm was configured with data type double, which means that it can handle decimal values for input parameters. The acceptance function was set to the default function, which accepts moves that result in a lower fitness value than the current state. The temperature function used was the temperature exp., which decreases the temperature exponentially during the optimization process. Combining the SA algorithm with the pattern search hybrid optimizer allowed us to refine the search space and find the optimal input parameters resulting in the desired results. Using these settings, we were able to obtain the best fitness value of 5.29212 for UTS, as shown in Fig. 10.

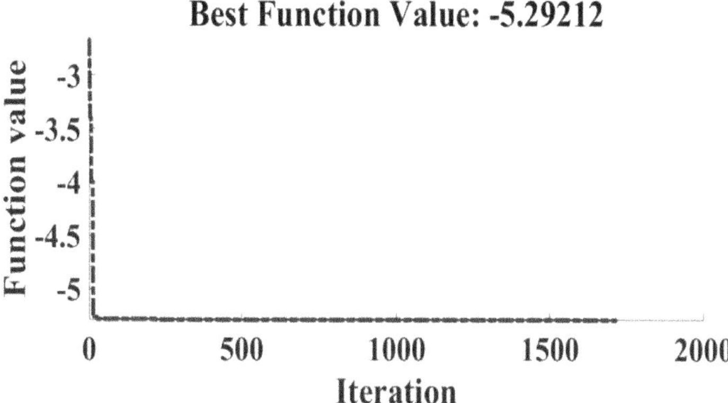

Figure 10: Simulated annealing: function value versus iteration plots

3.4 JAYA Algorithm Results

JAYA algorithm is a parameter less optimization algorithm that updates the population by finding better solutions in a cooperative manner. Despite not using any algorithm-specific parameters, JAYA algorithm has been shown to be highly effective in optimizing complex functions with multiple input parameters. Figure 11 depicts the function value versus iteration plot for JAYA algorithm.

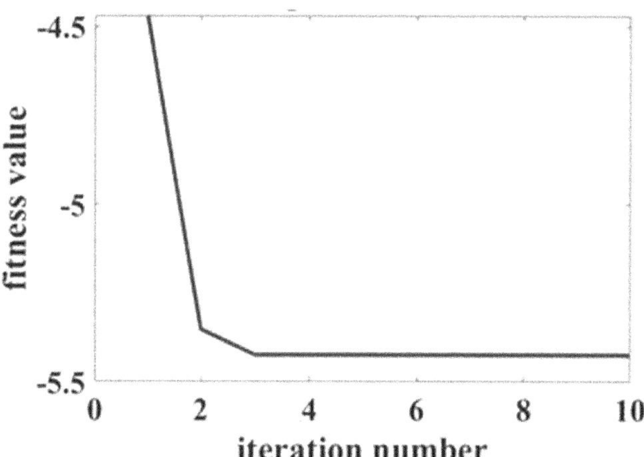

Figure 11: JAYA algorithm: function value versus iteration plot

3.5 CI Algorithm Results

CI is a method of optimization inspired by sociological principles developed by Kulkarni et al. [17]. Using this model, cohort candidates will achieve shared objectives in self-organizing systems. It is a group of candidates that interacts with each other and competes with each other. Essentially, the concept of a cohort is derived from the tendency of social individuals to follow, learn from, and adapt to the characteristics of others. Candidates in a cohort typically follow a specific behavioural pattern that may enhance their own behaviour. When a candidate attempts to follow a given behaviour characterized by particular qualities, they are likely to adopt these characteristics to improve their own behaviour. CI algorithm was investigated for mesh smoothing of complex objects (Sapre et al. [18]). CI algorithm is also used for solution of knapsack problems (Shabir and Kulkarni, [19]), numerical integration (Sapre et al. [20]) and discrete and mixed variable engineering problems (Kale and Kulkarni, [21]). Figure 12 depicts the function value versus iteration plot for CI algorithm.

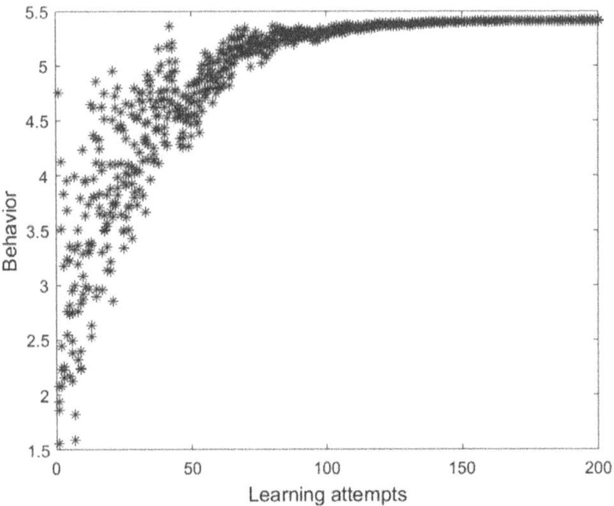

Figure 12: CI algorithm: behaviour versus learning attempts plot

3.6 Summary of Results

The CI algorithm, JAYA algorithm, GA, PSO, and SA algorithms were run for a 100 repetitions. Table 2 represents the results of these runs to understand the robustness of the algorithms. The robustness of the algorithm is a desirable quality in optimization applications, as it demonstrates the ability to provide consistent and reliable solutions to the problem at hand. The consistently similar results obtained from the CI algorithm, JAYA algorithm, GA, and PSO algorithms demonstrate their usefulness and effectiveness for solving optimization problems in various domains.

In the study, CI and JAYA algorithms outperformed other optimization algorithms considered to optimize UTS with multiple input parameters and achieved the best fitness value of 5.4242 MPa. This indicates that CI and JAYA algorithms are more efficient to obtain results. Overall, the results demonstrate the

Table 2: Summary table for algorithms robustness: function values for number of iterations

Algorithm	No. of runs	Time (sec)	Dwell time (sec)	Plunge depth (mm)	Spindle speed (rpm)	Fitness value (N/mm^2)	Standard deviation	Iterations
JAYA	100	0.12	40	1	1000	5.4242	0.02214	3
CI	100	12.24	40	1	1000	5.4242	0.03142	200
GA	100	1.91	40	1	1000	5.4242	0.09836	137
PSO	100	1.08	40	1	1000	5.4242	0.02421	22
SA	100	80.23	40.0012	1	1000.92	5.3112	0.67514	6324

Table 3: Validation test

Parameters	Optimal values	Predicted	Actual	Percentage error
Ultimate tensile strength (N/mm^2)	Spindle speed =1000 rpm, Plunge depth =1 mm, Dwell time = 40 sec	5.4242	5.5479 (average of five readings)	2.23

superiority of CI and JAYA algorithms over other popular optimization algorithms such as GA, PSO, and SA in terms of efficiency and effectiveness, especially when dealing with complex functions with multiple input parameters.

3.7 Validation Test

Table 3 displays the experimental validation of the model, which exhibited favourable conformity between the projected and factual outcomes. Upon executing the optimization process, further experiments were carried out to appraise the precision of the model. The optimized parameters obtained from the algorithms were implemented for these experiments, which corresponded to the most desirable outcomes. The UTS values resulting from these experiments were compared with the anticipated values from the algorithms, and the percentage error was computed.

4. Conclusions

The present study demonstrated that the optimization of the FSSW process to get the best possible weld strength using CI and JAYA algorithm during the welding of ABS material. CI and JAYA results were compared with GA, SA, and PSO results. The following are the key findings:

1. ABS material successfully welded by FSSW process. Dwell time was found to be the most significant parameter to achieve the best possible weld strength.
2. CI and JAYA algorithms performed better than GA, SA and PSO.
3. The robustness of the algorithm was checked by multiple runnings if the algorithms, which resulted in the same UTS values for all runs for the CI and JAYA algorithms.
4. Validation experiments confirmed that the obtained optimal input parameter settings by CI and JAYA algorithm resulted in UTS close to the predicted value of 2.23 percentage error.

References

1. Iftikhar, S.H., Mourad, A.H.I., Sheikh-Ahmad, J., Almaskari, F. and Vincent, S. 2021. A comprehensive review on optimal welding conditions for friction stir welding of thermoplastic polymers and their composites. *Polymers*, 13(8): 1208.
2. Aghajani Derazkola, H. and Simchi, A. 2018. Experimental and thermomechanical analysis of friction stir welding of poly(methyl methacrylate) sheets. *Science and Technology of Welding and Joining*, 23(3): 209–218.
3. Kumar, R., Singh, R. and Ahuja, I.P. 2019. Friction stir welding of ABS–15Al sheets by introducing compatible semi-consumable shoulder-less pin of PA6–50Al. *Measurement*, 131: 461–472.

4. Qiao, F., Cheng, K., Wang, L. and Guo, L. 2016. An experimental investigation on the dissimilar joining of AA6061 and 1Cr18Ni9Ti by refill friction stir spot welding and its mechanical properties. *Proceedings of the Institution of Mechanical Engineers, Part B: Journal of Engineering Manufacture*, 230(4): 779–785.
5. Taban, E., Gould, J.E. and Lippold, J.C. 2010. Dissimilar friction welding of 6061–T6 aluminum and AISI 1018 steel: Properties and microstructural characterization. *Materials & Design (1980-2015)*, 31(5): 2305–2311.
6. Yuan, W., Mishra, R.K., Webb, S., Chen, Y., Carlson, B., Herling, D. and Grant, G. 2011. Effect of tool design and process parameters on properties of Al alloy 6016 friction stir spot welds. *Journal of Materials Processing Technology*, 211: 972–977.
7. Yan, Y., Shen, Y., Zhang, W. and Hou, W. 2018. Friction stir spot welding ABS using triflute-pin tool: Effect of process parameters on joint morphology, dimension and mechanical property. *Journal of Manufacturing Processes*, 32: 269–279.
8. Adibeig, M.R., Hassanifard, S., Vakili-Tahami, F. and Hattel, J.H. 2018. Experimental investigation of tensile strength of friction stir welded butt joints on PMMA. *Materials Today Communications*, 17: 238–245.
9. Saeedy, S. and Besharati Givi, M.K. 2010. Experimental application of friction stir welding (FSW) on thermo plastic medium density polyethylene blanks. *In: Proceedings of the 10th Biennial Conference on Engineering Systems Design and Analysis*, (pp. 841–844). Istanbul, Turkey.
10. Moochani, A., Omidvar, H., Ghaffarian, S.R. and Goushegir, S.M. 2019. Friction stir welding of thermoplastics with a new heat-assisted tool design: Mechanical properties and microstructure. *Welding in the World*, 63: 181–190.
11. Azarsa, E. and Mostafapour, A. 2014. Experimental investigation on flexural behavior of friction stir welded high density polyethylene sheets. *Journal of Manufacturing Processes*, 16: 149–155.
12. Mosavvar, A., Azdast, T., Moradian, M. and Hasanzadeh, R. 2019. Tensile properties of friction stir welding of thermoplastic pipes based on a novel designed mechanism. *Welding in the World*, 63: 691–699.
13. Bilici, M.K., Yukler, A.I. and Kurtulmuş, M. 2011. The optimization of welding parameters for friction stir spot welding of high density polyethylene sheets. *Materials & Design*, 32: 4074–4079.
14. Bilici, M.K. 2012. Application of Taguchi approach to optimize friction stir spot welding parameters of polypropylene. *Materials & Design*, 35: 113–119.
15. Lambiase, F., Paoletti, A. and Di Ilio, A. 2015. Mechanical behaviour of friction stir spot welds of polycarbonate sheets. *International Journal of Advanced Manufacturing Technology*, 80: 301–314.
16. Bilici, M.K., Yukler, A.I., Kurtulmuş, M. and Kastan, A. 2018. The optimization of welding tool material and welding parameters in friction stir spot welding of plastics using Taguchi experimental design. *International Journal of Engineering Science and Applications*, 2: 47–53.
17. Kulkarni, A.J., Durugkar, I.P. and Kumar, M. 2013, October. Cohort intelligence: A self-supervised learning behaviour. *In: 2013 IEEE International Conference on Systems, Man, and Cybernetics*, pp. 1396–1400.
18. Sapre, M.S., Kulkarni, A.J. and Shinde, S.S. 2019. Finite element mesh smoothing using Cohort intelligence. *In: Proceedings of the 2nd International Conference on Data Engineering and Communication Technology* (pp. 469–480). Springer, Singapore.

19. Kulkarni, A.J. and Shabir, H. 2016. Solving 0–1 Knapsack problem using Cohort intelligence algorithm. *International Journal of Machine Learning and Cybernetics*, 7(3): 427–441.
20. Sapre, M.S., Kulkarni, A.J., Kale, I.R. and Pande, M.S. 2023. Application of Cohort intelligence algorithm for numerical integration. *In: Intelligent Systems and Applications* (pp. 445–453). Springer, Singapore.
21. Kale, I.R. and Kulkarni, A.J. 2018. Cohort Intelligence Algorithm for discrete and mixed variable engineering problems. *International Journal of Parallel, Emergent and Distributed Systems*, 33(6): 627–662.

CHAPTER

8

Supervised Machine Learning Based Classification of Dimensional Deviation of FDM 3D Printed Samples

Shreya Joshi[1], Rithvik Nair[1], Kaustubh Dwivedi[1], Bhargav Gadhiya[2], Mandar S. Sapre[1*], Ashwini V. Jatti[1]

[1] Symbiosis Institute of Technology, Symbiosis International (Deemed University), Pune, Maharashtra, India

[2] D.Y. Patil Institute of Technology, Savitribai Phule Pune University, Pune, Maharashtra, India

1. Introduction

Artificial intelligence's subfield of machine learning uses algorithms and statistical models to analyze data and generate predictions or judgements. Machine learning can be utilized in a variety of ways to enhance the effectiveness, efficiency, and dependability of the Fused Deposition Modelling (FDM) process.

Due to its accessibility, usability, and versatility, fused deposition modelling (FDM) printing is one of the most well-liked and frequently employed 3D printing methods. FDM printers are affordable for enthusiasts, small enterprises, and major manufacturers equally because they come in a variety of price ranges. It is a particular kind of 3D printing technique known as additive manufacturing. It produces a three-dimensional object by melting a thermoplastic filament and extruding it layer by layer.

Several studies have highlighted the significance of ML in optimizing AM processes. Colwell and Sankar (2019) explored the application of supervised ML for process control, demonstrating its potential for enhancing AM quality control [1]. Dey and Yodo (2019) conducted a systematic survey, emphasizing the role of ML in optimizing Fused Deposition Modelling (FDM) parameters [2].

*Corresponding author: mandar.sapre@sitpune.edu.in

Furthermore, Alafaghani et al. (2017) conducted experimental optimization of FDM processing parameters, employing a design-for-manufacturing approach. Their work underscores the practicality of using ML to fine-tune parameters for improved manufacturing outcomes [3].

Behkamal, Das, and McLeod (2021) presented a novel ML approach for quality assurance in metal AM. Their work demonstrated how ML can be harnessed to ensure product quality and reliability, a crucial concern in AM [4]. Liu et al. (2021) also explored the application of ML techniques for anomaly detection in AM, offering a valuable tool for identifying defects and ensuring product integrity [5].

Ding et al. (2021) introduced a multi-scale, ML-based framework that connects the AM process with resulting part structures. This approach facilitates a deeper understanding of the relationship between process parameters and final product properties, thereby enabling more precise control and optimization [6].

Shakeri et al. (2020) utilized grey relational analysis to optimize the mechanical strength and shape accuracy of polyamide FFF parts, demonstrating the potential of data-driven techniques in enhancing the mechanical performance of AM products [7]. Radhakrishnan et al. (2023) employed ML to predict feature dimensions for Fused Deposition Modelling, aiding in achieving greater dimensional accuracy [8].

Sheoran and Kumar (2020) conducted a comprehensive review of FDM process parameter optimization and its impact on mechanical properties and part quality. Their work offers insights into the ongoing research landscape in AM, highlighting the importance of parameter optimization [9].

Mahmood et al. (2018) implemented Taguchi-based process optimization for dimension and tolerance control in FDM. Their research exemplifies the practical application of statistical and ML techniques to improve part quality in AM [10].

Wuest et al. (2016) provided an overview of ML applications in manufacturing, including AM, emphasizing the advantages and challenges associated with implementing ML in the field [11]. Razvi et al. (2019) conducted a review of ML applications in AM, underscoring the growing interest and potential for innovation in this interdisciplinary area [12].

The structure of this essay makes it possible for the reader to swiftly comprehend and come to terms with the dependence on surface roughness. The introduction gives the reader a quick outline of the subject, and the literature review summarizes the most recent research in the field. The whole experimental setup and the used print parameters are then explained in the technique section. This section also provides a summary of each classification model as well as other crucial metrics that are used to gauge the validity of the findings. The classification results are analyzed, discussed, and the key takeaways are succinctly summed up in the results and comments section. Overall, the paper's structure makes sure that the reader can follow along and comprehend how the product's surface roughness is directly impacted by various printing conditions.

2. Methodology

2.1 Workflow

To have a consistent model, the ASTM E8 standard's geometry was chosen as the reference geometry. The dimensions were all evenly scaled back by 50% to lower the print size, and the thickness was set to 5 mm. Using the Design of Experiment (DoE) method and three levels of each input parameter, thirty alternative experiment conditions were created. Using UltimakerCura software, the CAD model was sliced under these test conditions and G-code was produced.

Using a Creality 3D FDM printer, the experimental inquiry was carried out. A different set of settings were applied to each print. The results of thirty prints with different layer heights, infill densities, infill patterns, bed temperatures, and nozzle temperatures were obtained. To structure the input parameters, a datasheet is then made. A digital vernier calliper was also used to calculate the changes in each model's length, width, and thickness from the original CAD file that we developed. Figure 1 illustrates the stepwise sequence for entire work. Figure 2a depicts the CAD drawing of specimen and Figure 2b shows the actual 3D printed specimens. Figure 2c shows the 3D printer used in present study.

The next step is to randomly divide these print parameters into two sets, each with twenty-four sets and six sets (with a 4:1 split), for the purposes of training and evaluating classification models, respectively. The machine learning application receives the first set (training set), which produces a correlation matrix that shows how dependent the input parameters are on one another as well as how dependent the supplied output parameters are on the input parameters.

The training and testing set of data were then subjected to classification models, and based on the outcomes of these models, performance evaluation was conducted. k-Nearest Neighbours (kNN Algorithm), Stochastic Gradient Descent (SGD) Algorithm and Kernel's Approximation Algorithm were used in this work as classification techniques. Overall, the study aimed to investigate the impact

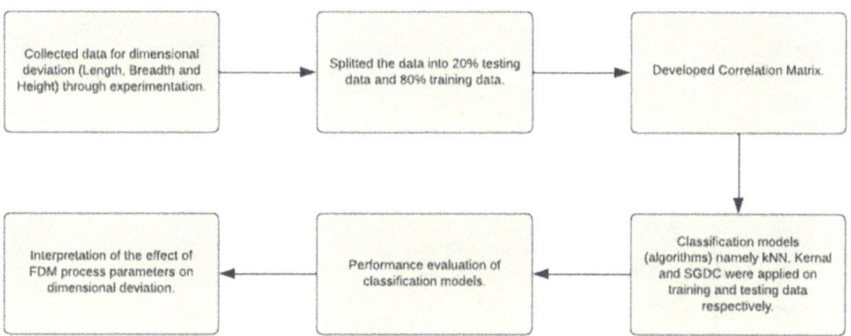

Figure 1: Workflow diagram

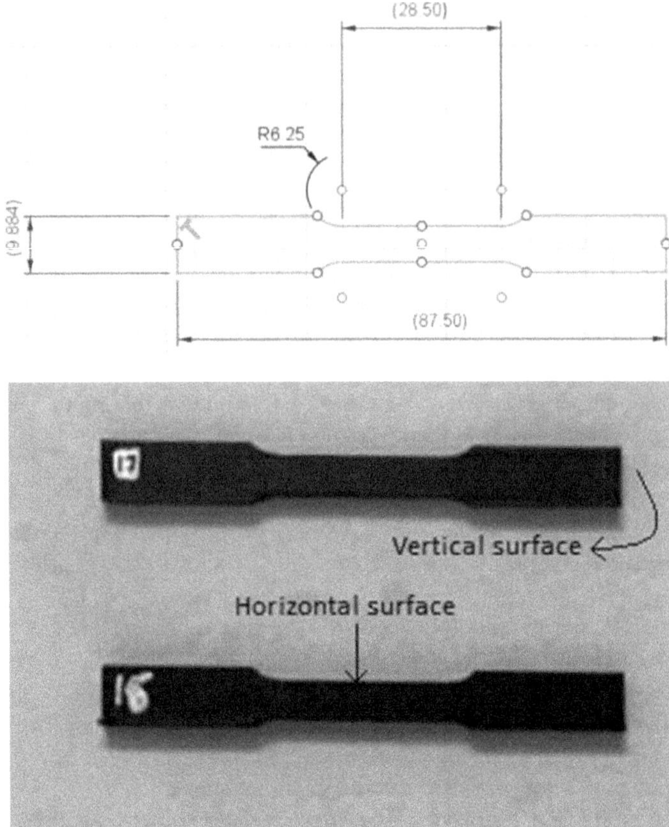

Figure 2(a) and (b): CAD drawing of the model and the final printed product. Note the markings making clear where the horizontal and vertical surfaces are. Classification models are used to predict surface roughness for both.

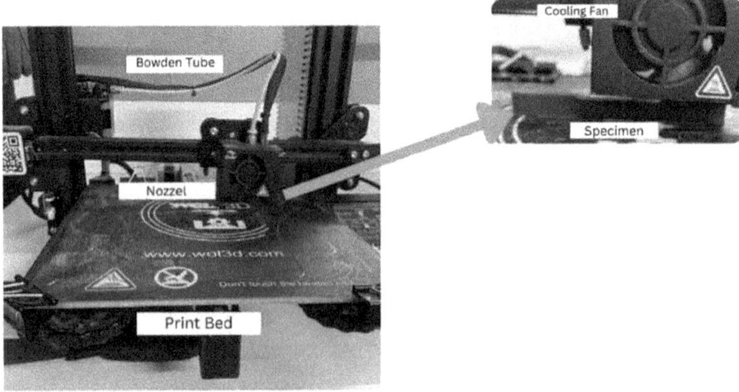

Figure 2(c): Creality 3D FDM printer

of FDM print parameters on the dimensional deviation, or in other words, how accurate the finished model is, by using these classification models. Based on the study's findings, the same print parameters can be optimized to produce a higher-quality print.

2.2 Background of the Classification Algorithms Applied

(i) k Nearest Neighbour

The k-Nearest Neighbours (kNN) algorithm is a popular classification algorithm in machine learning. The kNN algorithm converges to a new data point in the feature space while searching the k-nearest neighbours. A distance metric, such as Euclidean distance or Manhattan distance, is used to identify these k-nearest neighbours.

$$\text{Minkowski distance} = \left(\sum_{i=1}^{n} |x_i - y_i|\right)^{\frac{1}{p}} \tag{1}$$

A mathematical notion called the Minkowski distance is used to compare or contrast two points in a multidimensional space. It is a generalization of numerous other distance metrics, such as Manhattan and Euclidean distance, and it enables you to manage the degree of sensitivity to variations in various dimensions using a parameter called "*p*".FIn equation 1, x_i and y_i stand for the X and Y positions' respective coordinates in the *i*th dimension. A real number that is positive is the parameter *p*. This is used to calculate the Manhattan distance when it equals 1, and the Euclidean distance when it equals 2.

(ii) Stochastic Gradient Descent (SGD)

Stochastic Gradient Descent (SGD) is a variant of the Gradient Descent algorithm that is used for optimizing machine learning models. It addresses the computational inefficiency of traditional Gradient Descent methods when dealing with large datasets in machine learning projects.

In SGD, instead of using the entire dataset for each iteration, only a single random training example (or a small batch) is selected to calculate the gradient and update the model parameters. This random selection introduces randomness into the optimization process, hence the term "stochastic" in Stochastic Gradient Descent. Stochastic Gradient Descent algorithm:

- **Initialization:** Randomly initialize the parameters of the model.
- **Set Parameters:** Determine the number of iterations and the learning rate (alpha) for updating the parameters.
- **Stochastic Gradient Descent Loop:** Repeat the following steps until the model converges or reaches the maximum number of iterations.
- **Return Optimized Parameters:** Once the convergence criteria are met or

Table 1: Print parameters and calculated deviations

Trail condition	Layer height (mm)	Wall thickness (min)	Infill density (o/o)	Infill pattern	Nozzle temp. (°C)	Bed temp. (°C)	Print speed (mm/ sec)	Fan speed (%)	Lengthwise deviation	Breadth-wise deviation	Height-wise deviation
1	0.1	1	50	Honeycomb	200	60	120	0	0.34	0.1539	0.025
2	0.1	4	40	Grid	205	65	120	25	0.13	0.0589	-0.075
3	0.1	3	50	Honeycomb	210	70	120	50	0.28	0.0539	0.045
4	0.1	4	90	Grid	215	75	120	75	0.36	0.0839	0.045
5	0.1	1	30	Honeycomb	220	80	120	100	0.22	0.2039	-0.005
6	0.15	3	80	Honeycomb	200	60	60	0	0.2	0.2977	0.035
7	0.15	4	50	Grid	205	65	60	25	0.26	0.1389	0.005
8	0.15	10	30	Honeycomb	210	70	60	50	0.18	0.0964	0.025
9	0.15	6	40	Grid	215	75	60	75	0.36	0.2839	0.065
10	0.15	1	10	Honeycomb	220	80	60	100	0.15	0.1839	0.035
11	0.2	5	60	Honeycomb	200	60	40	0	0.12	0.4039	0.015
12	0.2	4	20	Grid	205	65	40	25	0.14	-0.0411	-0.025
13	0.2	5	60	Honeycomb	210	70	40	50	0.1	-0.0561	0.195
14	0.2	7	40	Grid	215	75	40	75	0.1	0.0614	0.015
15	0.2	3	60	Honeycomb	220	80	40	100	0.11	0.7789	0.015

16	0.1	1	50	Triangles	200	60	120	0	0.06	0.2614	0.025
17	0.1	4	40	Cubic	205	65	120	25	0.27	0.1364	0.135
18	0.1	3	50	Triangles	210	70	120	50	0.21	0.0339	0.025
19	0.1	4	90	Cubic	215	75	120	75	0.46	0.2014	0.075
20	0.1	1	30	Triangles	220	80	120	100	0.21	0.1814	0.045
21	0.15	3	80	Triangles	200	60	60	0	0.25	0.2039	0.035
22	0.15	4	50	Cubic	205	65	60	25	0.2	0.2364	-0.005
23	0.15	10	30	Triangles	210	70	60	50	0.25	0.2314	-0.085
24	0.15	6	40	Cubic	215	75	60	75	0.38	0.2364	0.065
25	0.15	1	10	Triangles	220	80	60	100	0.19	0.2964	0.035
26	0.2	5	60	Triangles	200	60	40	0	0.17	0.2039	-0.015
27	0.2	4	20	Cubic	205	65	40	25	0.14	0.2789	0.055
28	0.2	5	60	Triangles	210	70	40	50	0.27	0.4239	0.015
29	0.2	7	40	Cubic	215	75	40	75	0.21	0.3464	0.065
30	0.2	3	60	Triangles	220	80	40	100	0.08	0.0539	0.015

the maximum number of iterations is reached, return the optimized model parameters.

(iii) Kernal Approximation

Kernal Approximation is a technique used in machine learning to efficiently compute the results of Kernal-based algorithms, particularly Support Vector Machines (SVMs), on high-dimensional data. This approach is especially valuable when dealing with large datasets or complex Kernal functions. In this section, we'll explore the concept of Kernal Approximation and the algorithm associated with it.

The Kernal Approximation algorithm is designed to approximate the Kernal matrix without explicitly computing it. Instead, it maps the data to a lower-dimensional space, where the Kernal function can be efficiently applied. Here are the key steps of the Kernal Approximation algorithm:

- **Choose a Feature Map:** Select a suitable feature map that approximates the Kernal function. Popular choices include Random Fourier Features (RFF), Nystroem approximation, and Nyström-SVD.
- **Map Data:** Apply the chosen feature map to the training data and test data if applicable. This mapping reduces the data dimensionality, making it computationally tractable.
- **Train a Linear Model:** Train a linear classifier (e.g., Linear SVM or Logistic Regression) on the lower-dimensional feature representations of the training data.
- **Predictions:** To make predictions on new data points, apply the same feature map to map them into the lower-dimensional space and use the trained linear model for classification or regression.

2.3 Some Important Supervised Learning Terms

- **Precision:** Precision is a commonly used evaluation metric in machine learning, particularly for classification tasks. Precision measures the number of positive cases that were predicted to be positive. So, it measures the ratio between true positives (TP) to the actual positive predictions. The formula is given in equation 2.

  Precision for is defined as the ratio of number of true positives to the number of predicted positives.

$$\text{Precision} = \frac{\text{True Positive}}{\text{False Positive} + \text{True Positive}} \tag{2}$$

- **Recall (Sensitivity):** Recall is a commonly used evaluation metric in machine learning, particularly for classification tasks. Recall measures the number of positive cases that were correctly identified by the model. So, it measures the ratio between true positives (TP) to the actual positive cases. The formula is given in equation 3.

Recall is defined as the ratio of number of true positives to the total number of actual positives.

$$\text{Recall} = \frac{\text{True Positive}}{\text{False Negative} + \text{True Positive}} \tag{3}$$

- **F1 Score:** It gives a relative idea of Precision and Recall. It is maximum when Precision is equal to Recall. The formula is given in equation 4.

  F1 Score is the harmonic mean of precision and recall.

$$\text{F1} = 2.\frac{\text{Precision} \times \text{Recall}}{\text{Precision} + \text{Recall}} \tag{4}$$

- **AUC-ROC:** AUC-ROC stands for "Area Under the Receiver Operating Characteristic Curve". It is a metric commonly used in machine learning to evaluate the performance of a binary classification model, which predicts the probability of an input belonging to one of two classes. The ROC curve is a graphical representation of the performance of a binary classifier as its discrimination threshold is varied. It plots the true positive rate (sensitivity) against the false positive rate (1-specificity) at different threshold values. The

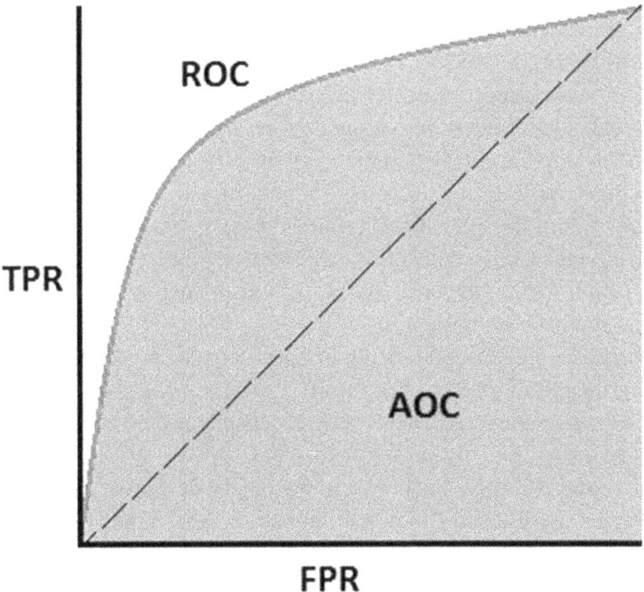

Figure 3: Receiver-Operator Characteristic (ROC)

AUC-ROC is the area under this curve, which ranges from 0 to 1, with a higher value indicating a better classifier performance. In essence, the AUC-ROC measures the ability of the model to distinguish between positive and negative classes, regardless of the chosen classification threshold. A perfect classifier has an AUC-ROC score of 1, while a random classifier has an AUC-ROC score of 0.5.

- **Specificity:** Specificity is another commonly used evaluation metric in machine learning, particularly for binary classification tasks. It measures the proportion of true negative (TN) predictions among all the actual negative cases in the dataset. The formula is given in equation 5.

  Specificity is defined as how many of the negative cases were correctly identified by the model.

$$\text{Specificity} = \frac{\text{True Negative}}{\text{True Negative} + \text{False Positive}} \tag{5}$$

3. Results and Discussions

3.1 Effect of FDM Process Parameters on Dimensional Deviation

The correlation matrix, also known as a heatmap, can be discovered using machine learning models, and this will enable us to read into the process parameters and discover the dependence of deviation on them. Understanding which process parameters can be altered and by how much they must be altered to obtain the desired dimensions is an important step because, in FDM, a flawless print requires striking the ideal balance between various process parameters.

Shown below are the correlation matrices for each length-wise, breadth-wise and height-wise deviations.

The correlation matrix for the length deviation is displayed in Fig. 4. It is noted that layer height, infill density, and print speed all play a significant role in this situation. By lowering the infill density and print speed and raising the layer height, we can produce accurate prints.

The correlation matrix for deviation in breadth is displayed in Fig. 5. As can be observed, print speed has some bearing. The accuracy of a print is improved by print speed. The layer height is also somewhat dependant. The accuracy increases as it decreases.

Figure 6 shows the correlation matrix for height deviation. As can be seen, infill density has a significant impact on height deviation. As the print size shrinks, the print accuracy grows.

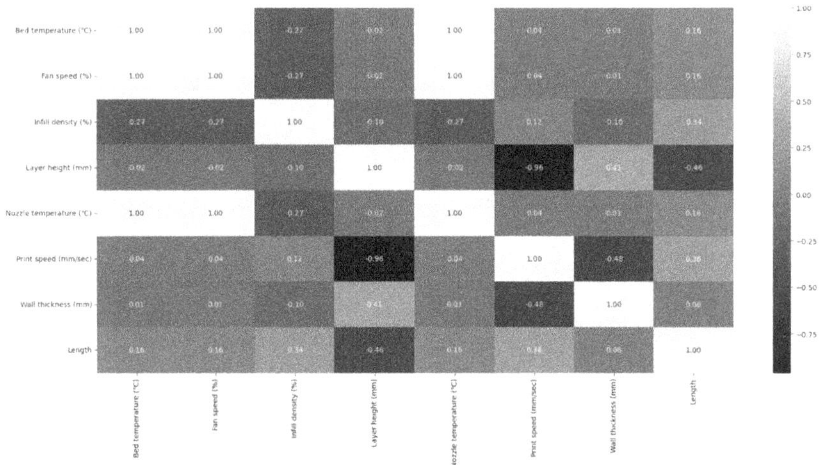

Figure 4: Correlation matrix for length-wise deviation

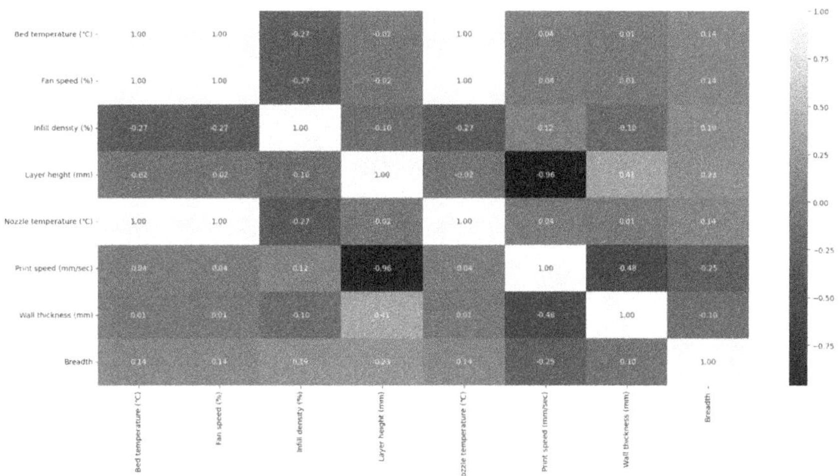

Figure 5: Correlation matrix for breadth-wise deviation

3.2 kNN Results

Length

The model has a moderate F1 score (0.55), indicating a reasonable balance between precision and recall. However, it has limited specificity (0.5), meaning it struggles to correctly identify negative cases. The AUC ROC (46.43%) suggests that the model's ability to distinguish between the two classes is only slightly better than random chance. Overall, there is room for improvement in the model's performance, especially in correctly classifying negative cases. Figure 7(a)

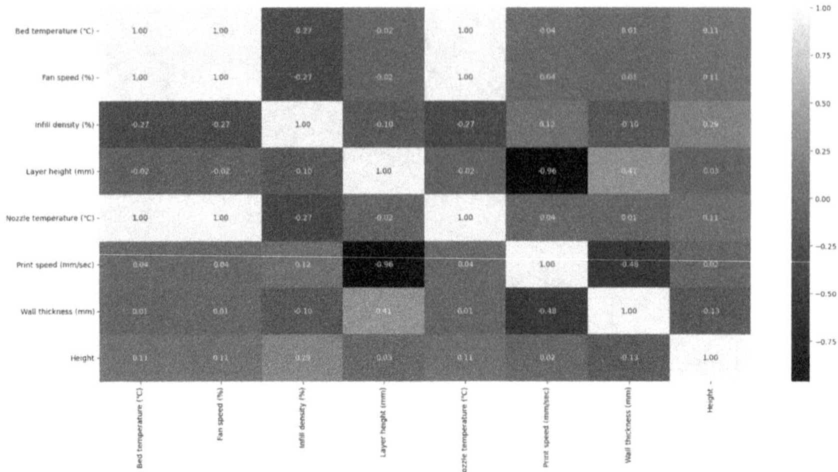

Figure 6: Correlation matrix for height-wise deviation

and (b) depicts the Confusion matrix and ROC for kNN model for length-wise deviation, respectively.

Breadth

In brief, the model's F1 score of 0.5 indicates a fair balance between precision and recall. However, the specificity of 0.5 suggests that it performs no better than random guessing in identifying negative cases. The AUC ROC of 58.33% indicates a slightly better-than-chance ability to distinguish between positive and negative cases. Overall, there is room for improvement in the model's performance, especially in correctly classifying both positive and negative cases. Figure 8(a) and (b) depicts the Confusion matrix and ROC for kNN model for breadth-wise deviation, respectively.

Height

In brief, the model's F1 score of 0.67 indicates a reasonably balanced performance between precision and recall. The specificity of 0.6 suggests that it is relatively effective at correctly identifying negative cases. The AUC ROC of 67.5% indicates a good ability to distinguish between positive and negative cases. Overall, the model shows a promising performance with room for potential improvement. Figure 9(a) and (b) depicts the Confusion matrix and ROC for kNN model for height-wise deviation, respectively.

3.3 Kernal Approximation Results

Length

In brief, the model's F1 score of 0.44 indicates a moderate balance between

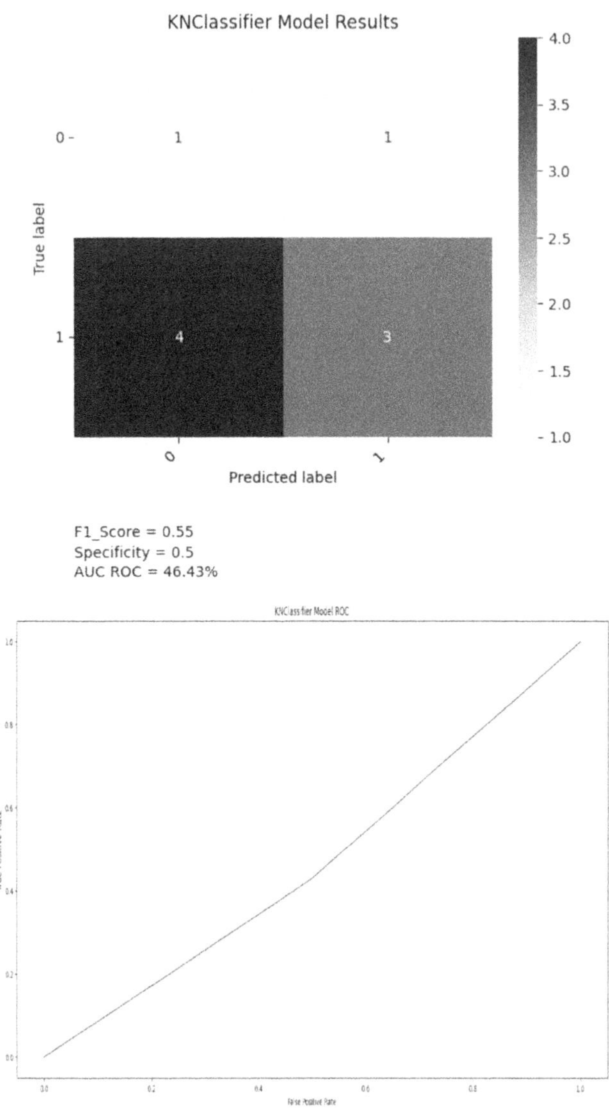

Figure 7(a) and (b): Confusion matrix and ROC for kNN model for length-wise deviation

precision and recall, with room for improvement. The specificity of 1 suggests it performs very well in identifying negative cases. The AUC ROC of 64.29% indicates a reasonable ability to distinguish between positive and negative cases. Overall, the model is strong at classifying negative cases but may need enhancement in correctly identifying positive cases for better overall performance. Figure 10(a) and (b) depicts the Confusion matrix and ROC for Kernal Approximation model for length-wise deviation, respectively.

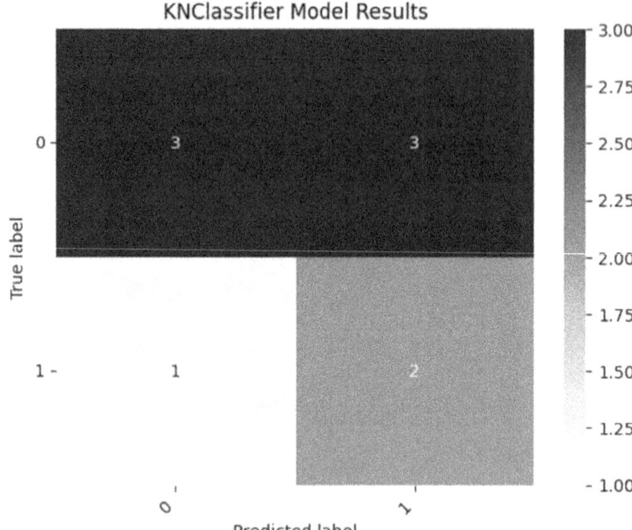

F1_Score = 0.5
Specificity = 0.5
AUC ROC = 58.33%

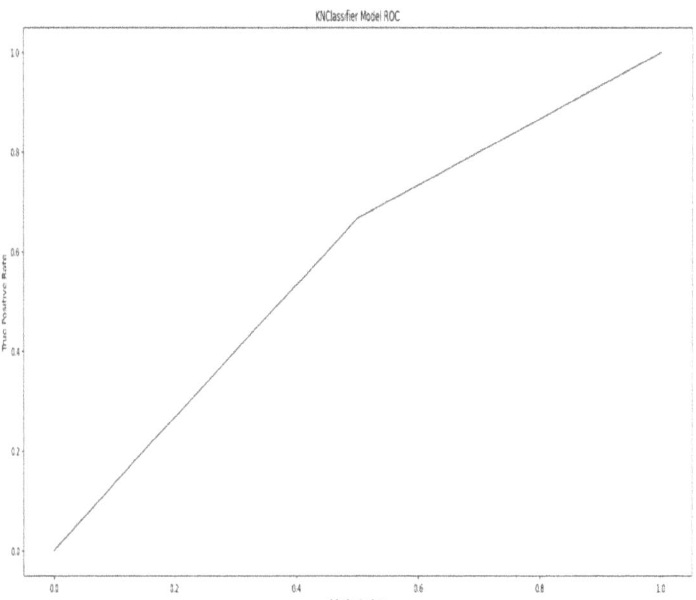

Figure 8(a) and (b): Confusion matrix and ROC for kNN model for breadth-wise deviation

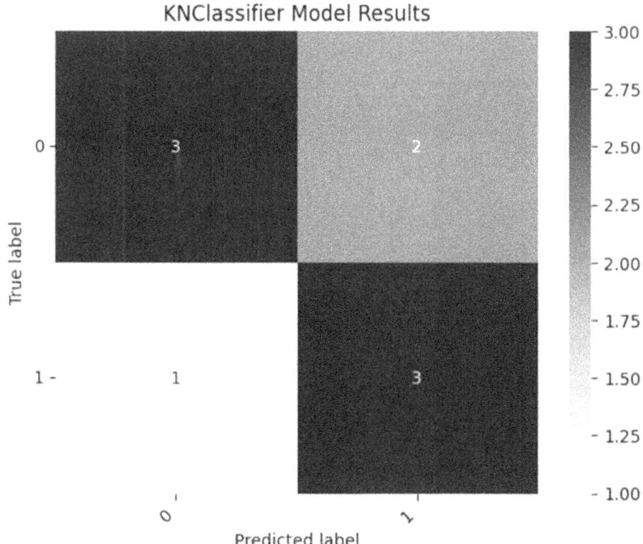

F1_Score = 0.67
Specificity = 0.6
AUC ROC = 67.50%

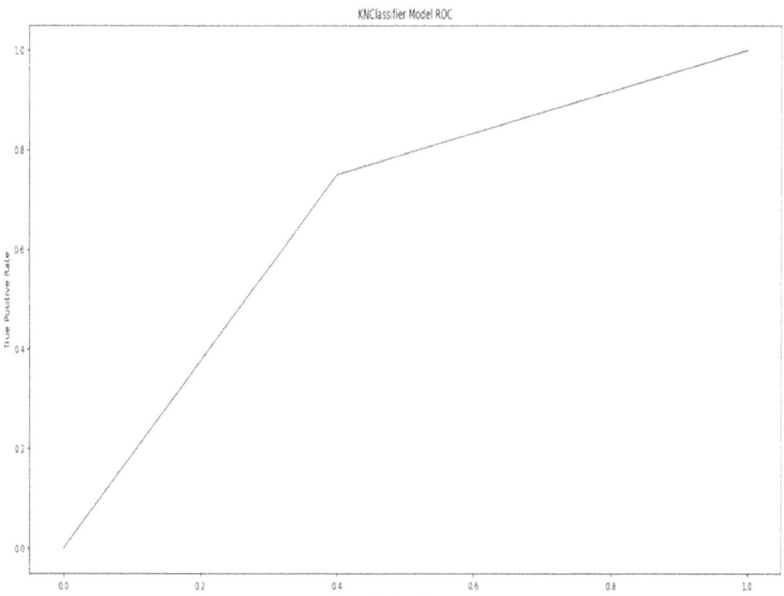

Figure 9(a) and (b): Confusion matrix and ROC for kNN model for height-wise deviation

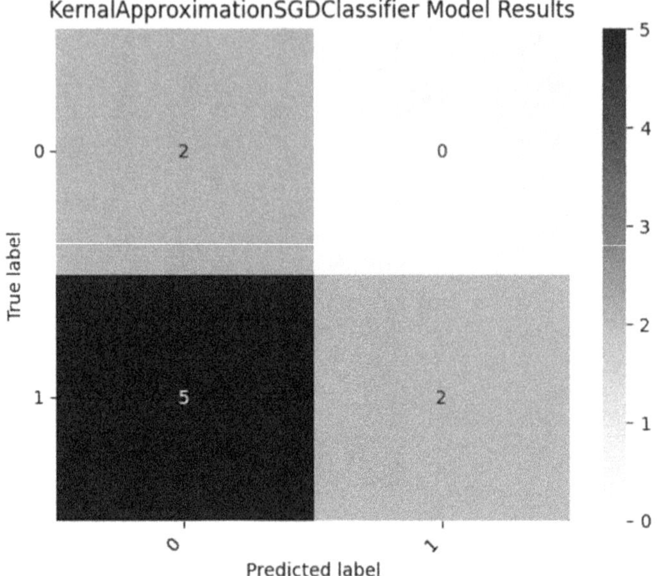

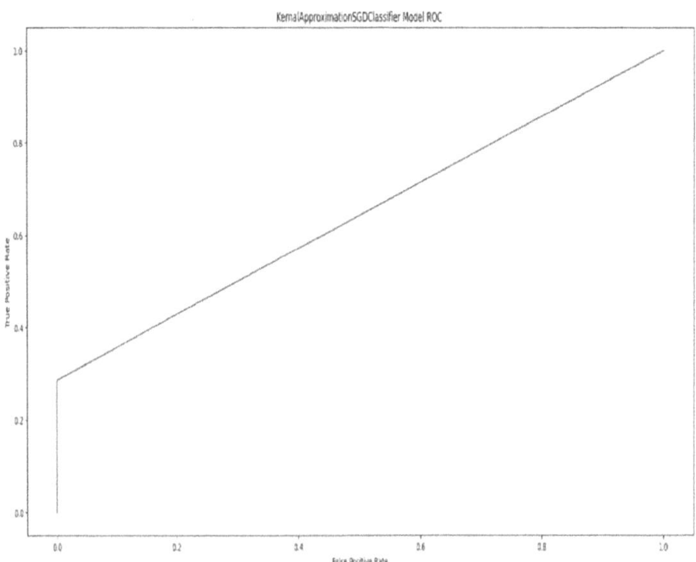

Figure 10(a) and (b): Confusion matrix and ROC for Kernal Approximation model for length-wise deviation

Breadth

In brief, the model's F1 score of 0.44 suggests a moderate balance between precision and recall. The specificity of 0.33 indicates that it struggles to correctly identify negative cases. The AUC ROC of 50.00% implies that the model's ability to distinguish between positive and negative cases is no better than random guessing. Overall, the model's performance is subpar and requires significant improvement, especially in correctly classifying negative cases. Figure 11(a) and (b) depicts the Confusion matrix and ROC for Kernal Approximation model for breadth-wise deviation, respectively.

Height

In brief, the model's F1 score of 0.67 indicates a reasonably balanced performance between precision and recall. The specificity of 0.6 suggests that it is relatively effective at correctly identifying negative cases. The AUC ROC of 67.5% indicates a good ability to distinguish between positive and negative cases. Overall, the model shows a promising performance with a solid ability to classify both positive and negative cases. Figure 12(a) and (b) depicts the Confusion matrix and ROC for Kernal Approximation model for height-wise deviation, respectively.

3.4 Stochastic Gradient Descent Results

Length

In brief, the model's F1 score of 0.67 indicates a reasonably balanced performance between precision and recall. The specificity of 0.5 suggests that it performs no better than random guessing in identifying negative cases. The AUC ROC of 53.57% implies a limited ability to distinguish between positive and negative cases. Overall, the model demonstrates moderate effectiveness but may need improvement, particularly in correctly classifying negative cases and overall discriminatory power. Figure 13(a) and (b) depicts the Confusion matrix and ROC for SGD model for length-wise deviation, respectively.

Breadth

In brief, the model's F1 score of 0.67 indicates a reasonably balanced performance between precision and recall. The high specificity of 0.833 suggests it is quite effective at correctly identifying negative cases. The AUC ROC of 75% signifies a good ability to distinguish between positive and negative cases. Overall, the model exhibits strong performance with a good ability to classify both positive and negative cases. Figure 14(a) and (b) depicts the Confusion matrix and ROC for SGD model for breadth-wise deviation, respectively.

Height

In brief, the model's F1 score of 0.8 indicates a strong balance between precision

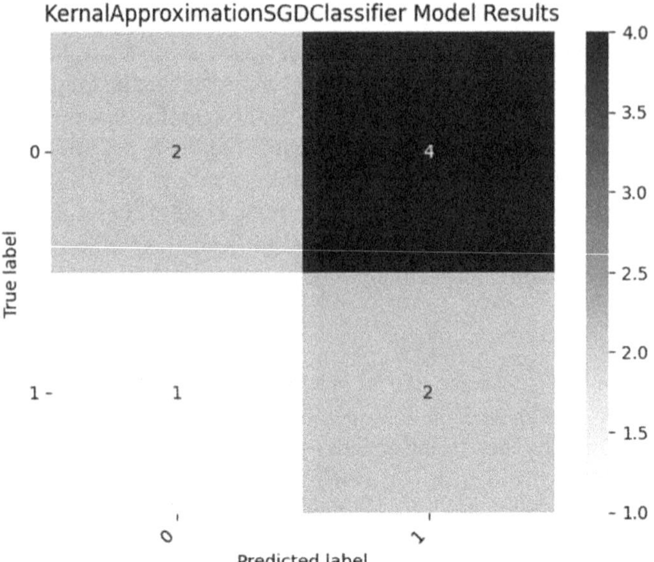

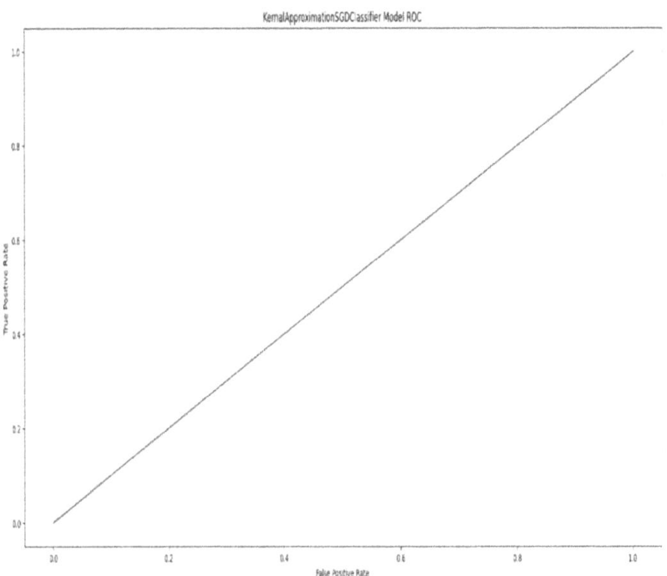

Figure 11(a) and (b): Confusion matrix and ROC for Kernal Approximation Model for breadth-wise deviation

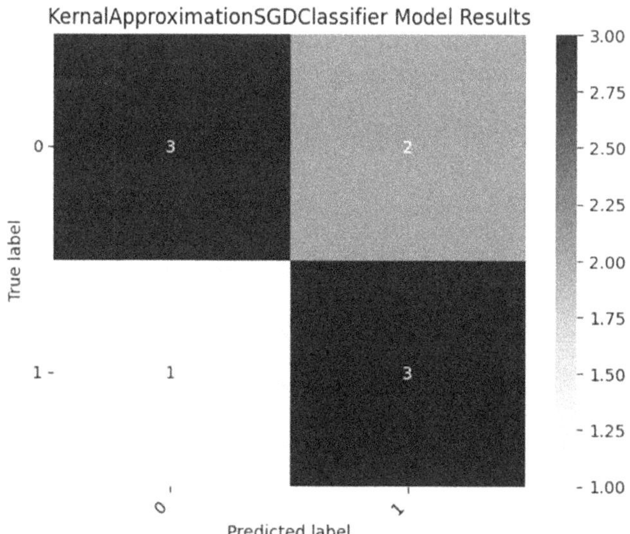

F1_Score = 0.67
Specificity = 0.6
AUC ROC = 67.50%

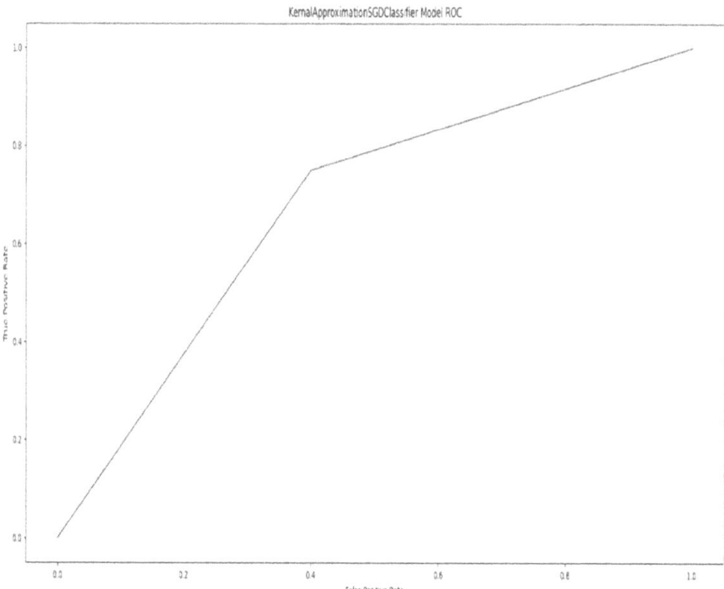

Figure 12(a) and (b): Confusion matrix and ROC for Kernal Approximation model for height-wise deviation

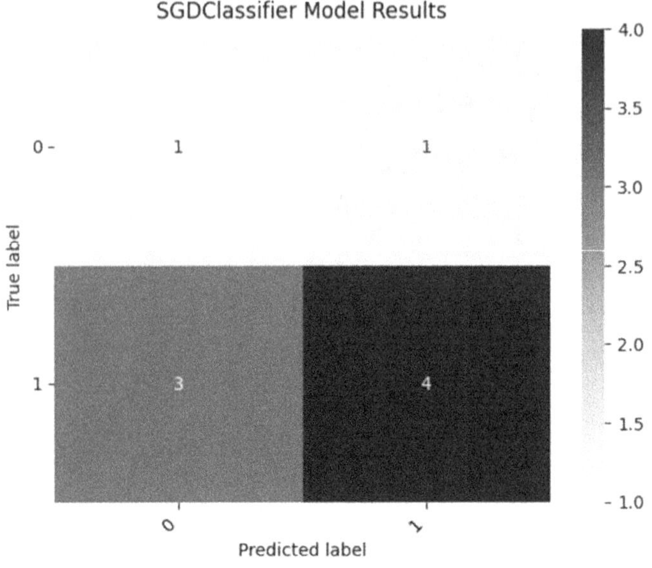

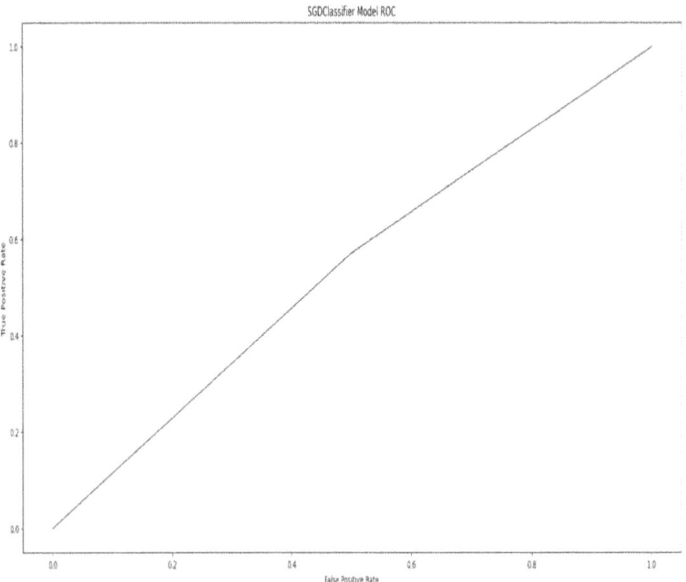

Figure 13(a) and (b): Confusion matrix and ROC for SGD model for length-wise deviation

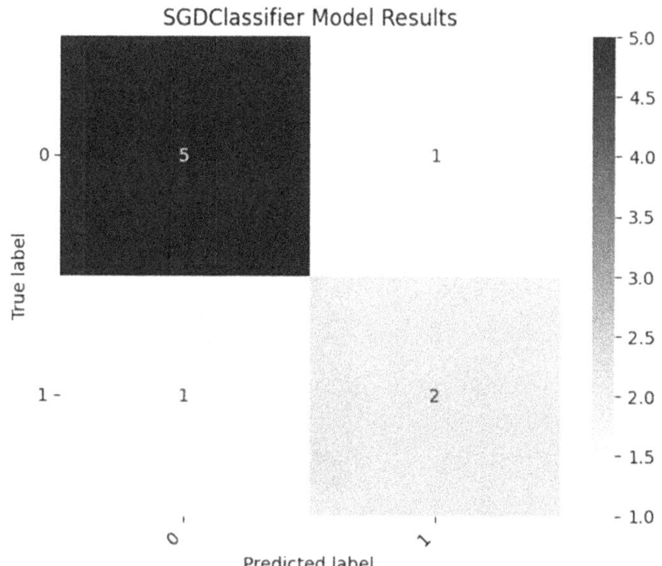

F1_Score = 0.67
Specificity = 0.8333333333333334
AUC ROC = 75.00%

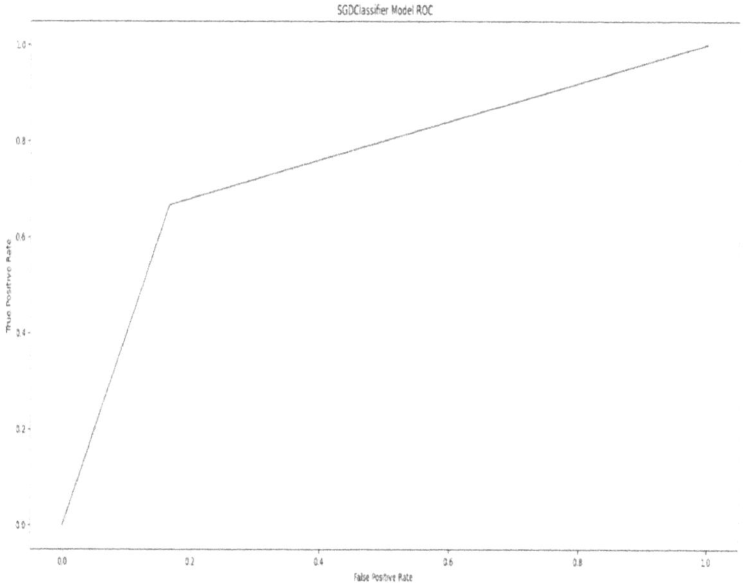

Figure 14(a) and (b): Confusion matrix and ROC for SGD model for breadth-wise deviation

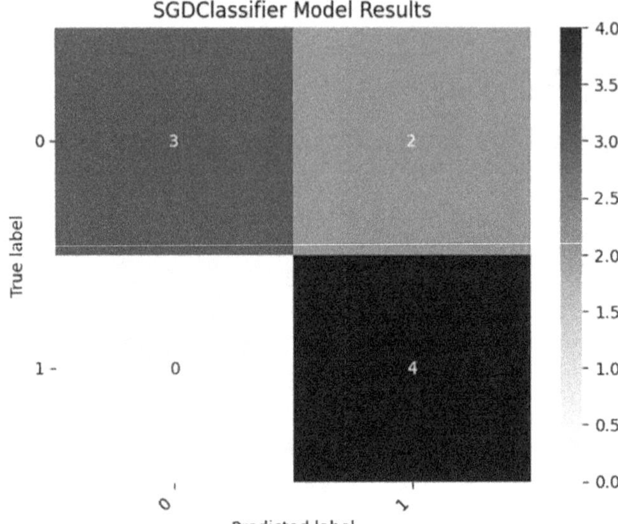

F1_Score = 0.8
Specificity = 0.6
AUC ROC = 80.00%

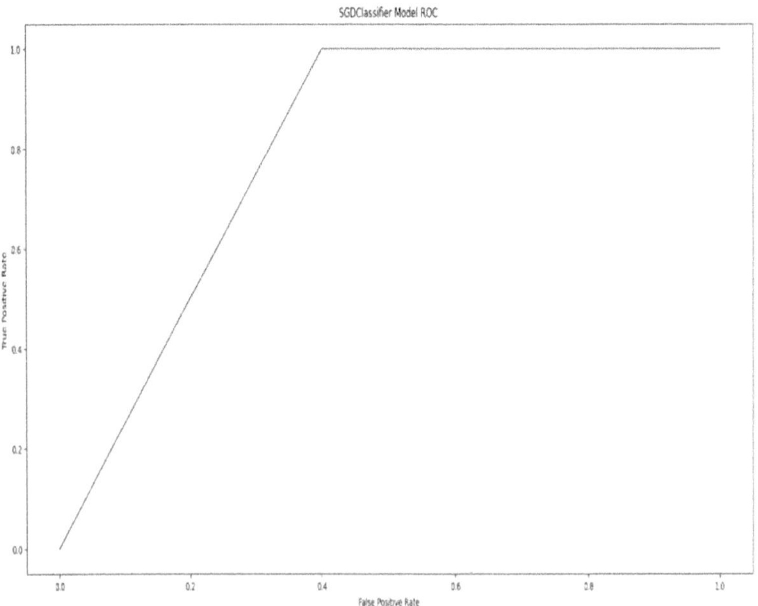

Figure 15(a) and (b): Confusion matrix and ROC for SGD model for height-wise deviation

and recall. The specificity of 0.6 suggests it is moderately effective at correctly identifying negative cases. The AUC ROC of 80% signifies a very good ability to distinguish between positive and negative cases. Overall, the model demonstrates excellent performance with a strong ability to classify both positive and negative cases. Figure 15(a) and (b) depicts the Confusion matrix and ROC for SGD model for height-wise deviation, respectively.

4. Conclusion

The present study employed supervised machine learning classifiers namely, k-Nearest Neighbours, Kernel Approximation and Stochastic Gradient Descent to predict dimensional deviation of 3D printed parts. Based on the results of the study,it was found that layer height is the most significant parameter that affects length-wise and breadth-wise deviation and infill density is the most significant parameter that affects height-wise deviation. Based on the values of the F1-score and AUC ROC, it can be inferred that for any given deviation, Kernel Approximation and Stochastic Gradient Descent outperformed the k-Nearest Neighbours Algorithm. As a result, it can be concluded that supervised machine learning classifier algorithms are suitable for use in 3D printing manufacturing.

References

1. Liu, J., Guo, Q., Xiong, W., Guo, Y. and Hu, X. 2021. Anomaly detection for additive manufacturing using machine learning techniques. *Robotics and Computer-Integrated Manufacturing*, 69: 101979.
2. Ding, J., Zhang, D., Wang, D., Chen, D. and Tang, H. 2021. A multi-scale, machine learning-based process-structure linkage framework for additive manufacturing. *Journal of Manufacturing Processes*, 65: 1–11.
3. Colwell, M.J. and Sankar, B.V. 2019. Additive manufacturing process control with supervised machine learning. *Materials Science and Engineering: A*, 742: 170–176.
4. Behkamal, B., Das, O. and McLeod, R.R. 2021. Machine learning approach to quality assurance in metal additive manufacturing. *Additive Manufacturing*, 38: 101814.
5. Shakeri, Z., Benfriha, K., Zirak, N. and Shirinbayan, M. 2020. Mechanical strength and shape accuracy optimization of polyamide FFF parts using grey relational analysis. *Materials Today: Proceedings*, 33(2020): 2674–2678.
6. Sheoran, A.J. and Kumar, H. 2020. Fused deposition modeling process parameters optimization and effect on mechanical properties and part quality: Review and reflection on present research. *Materials Today, Proceedings*, ISSN 2214–7853.
7. Dey, A. and Yodo, N. 2019. A systematic survey of FDM process parameter optimization and their influence on part characteristics. *Journal of Manufacturing and Materials Processing*, 3(3): 64.

8. Radhakrishnan, L., Rajendran, P., Biswal, R., Goswami, A.G. and Ganesan, A. 2023. Machine learning approach to predict feature dimensions for fused deposition modelling. *Transactions of the Indian Institute of Metals*, 76: 315–321.
9. Shahrain Mahmood, A.J. and Qureshi, D.T. 2018. Taguchi based process optimization for dimension and tolerance control for fused deposition modelling. *Additive Manufacturing*, 21, ISSN 2214–8604.
10. Wuest, T., Weimer, D., Irgens, C. and Thoben, K.D. 2016. Machine learning in manufacturing: Advantages, challenges, and applications. *Production & Manufacturing Research*, 4(1): 23–45.
11. Razvi, S.S., Feng, S., Narayanan, A., Lee, Y.T. and Witherell, P. 2019. A review of machine learning applications in additive manufacturing. *Proceedings of the ASME 2019 International Design Engineering Technical Conferences and 39th Computers and Information in Engineering Conference*. Vol. 1. Anaheim, California, USA.
12. Alafaghani, A., Qattawi, A., Alrawi, B. and Guzman, A. 2017. Experimental optimization of fused deposition modelling processing parameters: A design-for-manufacturing approach. *Procedia Manufacturing*, 10, ISSN 2351–9789.

CHAPTER

9

Polymer Composite Flexural Strength Estimation using K-Nearest Neighbouring Classification Algorithm

Dhruv A. Sawant[1*], Ashwini V. Jatti[2]

[1] Department of Mechanical Engineering, Symbiosis Institute of Technology, Symbiosis International (Deemed University), Pune, India

[2] Department of Instrumentation Engineering, Dr. D.Y. Patil Institute of Technology, Savitribai Phule Pune University, Pune, India

1. Introduction

In today's highly advanced industrial climate, manufacturers from all over the world strive to maintain their dominant position in the worldwide market. This attempt to maintain their position tries to generate every conceivable discipline of technology. The complete prototype's development and design laid the foundation for the production of the finished products. Polymer matrix composite materials are currently displacing fibre plastic, metals, and alloys as the primary raw materials for manufacturing. Given the foregoing, the current study's goal was to develop a novel polymer composite matrix material with outstanding mechanical properties that could be trustworthy and meet the rising demand from the manufacturing industries on the international market. Natural fibre's intriguing qualities, including its renewability and biodegradability, make it possible for it to be used in military vehicles, textiles, biomedical applications, structural and architectural constructions, and the automobile and aerospace sectors [1]. Natural fibres are starting to replace glass fibres in composite materials as they are less expensive, lighter, and appear to be more environmentally friendly. The important factors influencing the relative environmental performance of natural fibre and

*Corresponding author: dhruv.sawant.btech2022@sitpune.edu.in

glass fibre composites are reviewed in a few comparative life cycle assessment studies [2]. The interactive effects of process factors (drill diameter, feed, and cutting speed) on the delamination factor and surface polish were examined using the Box-Behnken Design method [3]. Quality of life for patients can be improved by using biocomposites to replace or act as a framework for the regeneration of injured, degraded tissues and organs [4]. Various (non-)renewable engineering/commodity polymers and commercially available randomly oriented glass fibre-reinforced polymers (GFRP) are used to compare the tensile properties of these composite materials [5]. Usually, metallic or halogenated polymeric coatings are needed to improve the gas barrier; however, both have negative environmental effects, and metallic coatings restrict electromagnetic and visible light [6]. Fibre-reinforced composites' flexural characteristics are significantly influenced by fibre factor. Theoretically, long fibres combined with strong fibre-matrix bonds can result in high composite strength, while short fibres affect the composite's ductility [7]. We examined how hybridization affected the flexural strength and modulus of injection-molded ABS polymers with single-gated (SG) and double-gated (DG) reinforcements of both short glass fibres (GF) and spherical glass beads (GB). Flexural strength and the modulus of SG and DG ABS/GF/GB hybrids were shown to rise with the total concentration of glass and the concentration of glass fibres in the hybrid both increased [8]. Micromechanical modelling of composites made of polypropylene (PP) was used to determine Young's modulus of henequen fibres, and when a correction procedure was applied, a single filament tensile test was used to further confirm the estimate [9]. The fabrication of composite materials from bio-polyethylene reinforced with maize stover fibres has been explored with the intention of improving sustainability, contributing to the circular economy, and combating climate change [10]. Micro droplet experiments were used to examine and compare the effects of three maleic anhydride-modified polypropylene (MAPP) specimens with varied MA concentrations and crystallinities on the interfacial shear strengths (IFSSs) of PP and carbon fibre (CF) [11]. To investigate the mechanical behaviour of the hybrid composites, static tensile tests were run. The rule of hybrid mixtures and Halpin-Tsai equations were used to explore the hybrid effect on the elastic modulus of the composites [12]. Natural fibres from lignocellulosic materials have been used to create green materials in woven form recently for a variety of uses, including flooring, ballistic materials, household appliances, car parts, aerospace components, and structural and non-structural composites [13]. Recycled newspaper fibres (ONP) have great specific characteristics and have the potential to be superb nonabrasive reinforcing fibres [14]. The addition of mechanical wood pulp to the polypropylene matrix resulted in a noticeable nucleating impact by raising both the temperature at which the polymer degrades and its degree of crystallinity [15].

The flexural strength of various ABS-Cu and ABS-Al composition variations is examined in the current work. The paper is divided into three sections: the tools and procedures for testing the flexural resistance of flexural specimens; the generalization of machine learning classification; the results; and discussions of the experimental data recorded; and the analysis of the experimental data of ABS-Cu and ABS-Al using machine learning classification.

2. Materials and Methods

In a flexure test, the specimen's convex side experiences tensile stress while the concave side experiences compression stress. Because of this, there is shear tension along the midline. Shear stress must be kept to a minimum to guarantee that tensile or compression stress will be the main cause of failure. Specimens were chosen in accordance with ASTM D790 standards. The specimens were checked to make sure there were no cracks. Steps were taken to verify the computerized system's proper connectivity as well as the required input setup details, such as the specimen's gauge length and gauge diameter. Only then will the system think about releasing the machine's captured data. As the test is conducted on the tensile specimen of the specified standard, the computerized system displays data visually on a digital screen that is electronically recorded by the machine.

Figure 1(a): ABS-Cu flexural specimen

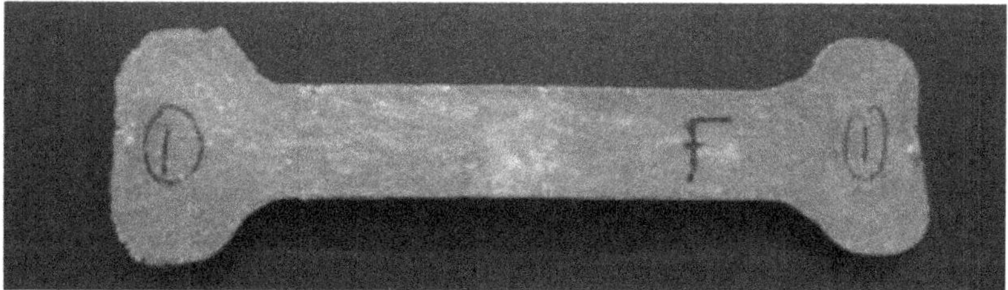

Figure 1(b): ABS-Al flexural specimen

3. Machine Learning

This work combined sample flexural testing with a range of machine-learning classification algorithms for different material compositions (including copper, aluminum, ABS, and surfactant substances).

In order to accurately classify the flexural strength and display confusion matrices and AUC-ROC curves for further analysis, sample data were generated using the synthetic data creation process in Matlab based on the experimental data. Fresh observations are classified using training data via the Classification algorithm, a Supervised Learning technique. In classification, a program learns how to categorize new observations into various groups or classes using the dataset or provided observations. In the current study, the value above the average of the flexural strength values is treated as 1, while the value below it is regarded as 0. This divides the data into two distinct groups. The current study employed the k-nearest neighbour (kNN) classification technique, which determines the shortest Euclidean distance between the average value and each value of flexural strength.

1. The number of neighbours hyperparameter option controls the number of nearest neighbours in the target dataset to be used in the kNN classification algorithm to classify each value or point.
2. The distance metric, this metric determines how far apart two locations are from one another.
3. Whether a distance is equal or inverse is determined by the distance weight (1/distance).

The confusion matrix was visualized using the metrics module of the Python sklearn package. The dataset was split into 80% training data and 20% random test data in order to acquire accurate findings for the data prediction. While the False positives (FP) and False negatives (FN) in the confusion matrix reflect wrongly predicted data or point out an error in the data prediction, the True positives (TP) and True negatives (TN) in the confusion matrix represent properly predicted data. To collect a wide range of categorization data, only the values of k=1, 2, 3, and 5 were considered for the number of neighbours (k). Performance for categorization issues at various threshold levels is measured by the AUC-ROC curve. AUC stands for the level or measurement of separability, and ROC is a probability curve. It illustrates how well the model is in differentiating between classes. The model performs better at classifying the 0 class as 0, and the 1 class as 1, the higher the AUC.

Table 1(a): Flexural strength for ABS-Cu composition

Sr no.	ABS %	Copper %	Surfactant material %	Flexural
1	44.05	50.18	5.77	47.96
2	23.03	69.87	6.96	32.08
3	65.36	30.28	4.87	42.46
4	...	...	...	...
...	...	...	...	...
100	43.83	50.06	5.99	47.94

Table 1(b): Flexural strength for ABS-Al composition

Sr no.	ABS%	Aluminum%	Surfactant material %	Flexural
1	44.05	50.18	5.77	55.52
2	23.03	69.87	6.96	28.22
3	65.36	30.28	4.87	56.47
4	...	...	...	...
...	...	...	...	...
100	43.83	50.06	5.99	55.50

4. Results and Discussions

According to the graph in Fig. 2(a), the composition ABS-Cu with ABS percentage equal to 44.05, copper percentage equal to 50.18, and surfactant material percentage equal to 5.77 had the highest flexural strength, which was 47.96. This demonstrates how flexural strength diminishes as copper content rises. The ABS-Al composition had a maximum flexural strength of 62.78, measured 6.75, and contained 65.07 percent ABS, 29.88 percent aluminum, and 5.07 percent surfactant material. This demonstrates that when the amount of aluminum in the polymer matrix increases, the flexural strength falls. Figures 4(a), 4(b), and 4(c)

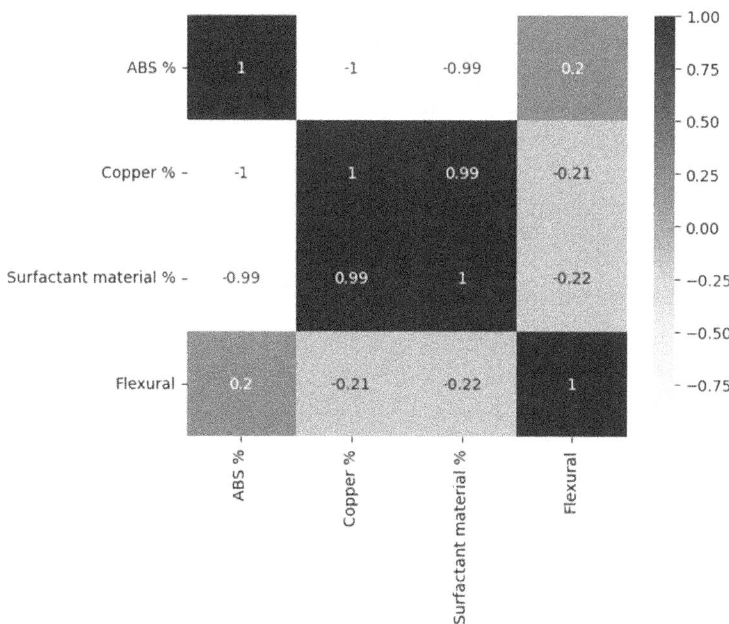

Figure 2(a): ABS-Cu composition heatmap

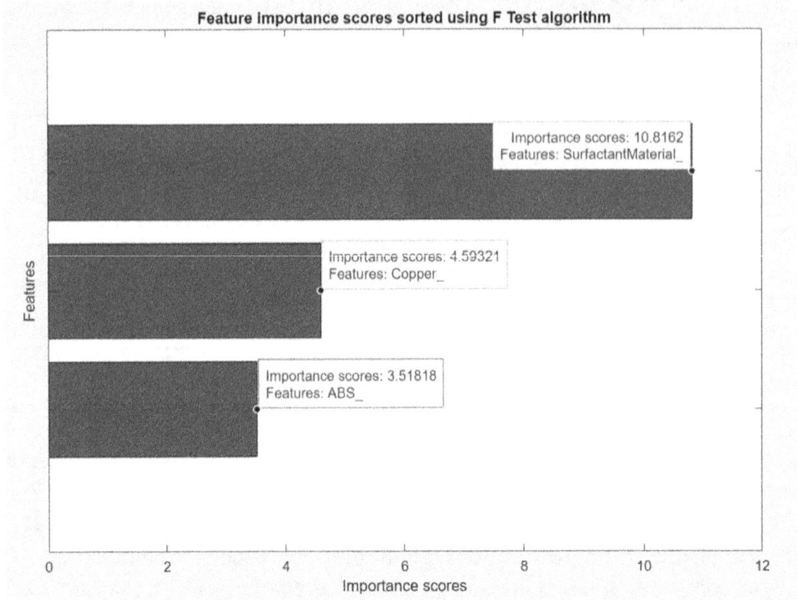

Figure 2(b): F-test for ABS-Cu feature selection

and Figs. 5(a), 5(b), and 5(c). provide data from ABS-Cu and ABS-Al composite flexural testing, respectively. The heatmap in Fig. 2(a). shows how ABS, copper, surfactant, and the flexural strength of an ABS-Cu composite material are related. According to the F-test results, copper and ABS have a very small impact on the flexural strength of the composites. Surfactant material composition, on the other hand, strongly influences flexural strength.

According to Figs. 3(a) and 3(b), which depict the ABS-Al F-test and heatmap correlation matrix, aluminum% has the greatest impact on the composition of ABS-Al's flexural strength, followed by ABS% and surfactant material%.

4.1 Machine Learning Classification: ABS-Cu

For the current investigation, the k-nearest neighbour classification accuracy for flexural strength of ABS-Cu composition for k=1, 2, 3 and 5. For the flexural strength of ABS-Cu, the composition was 75%, 80%, 75%, and 75%, respectively. The Euclidean distance between the nearest and average values of flexural strength is calculated using the k-nearest neighbour method. The values that were true positives (1) were considered to be flexural strength values above the average value which was 40.05 and true negatives (0) were considered below the average value. The confusion matrix is divided into 4 quadrants for study.

1. Top-left quadrant: True negative
2. False Positive in the top-right quadrant

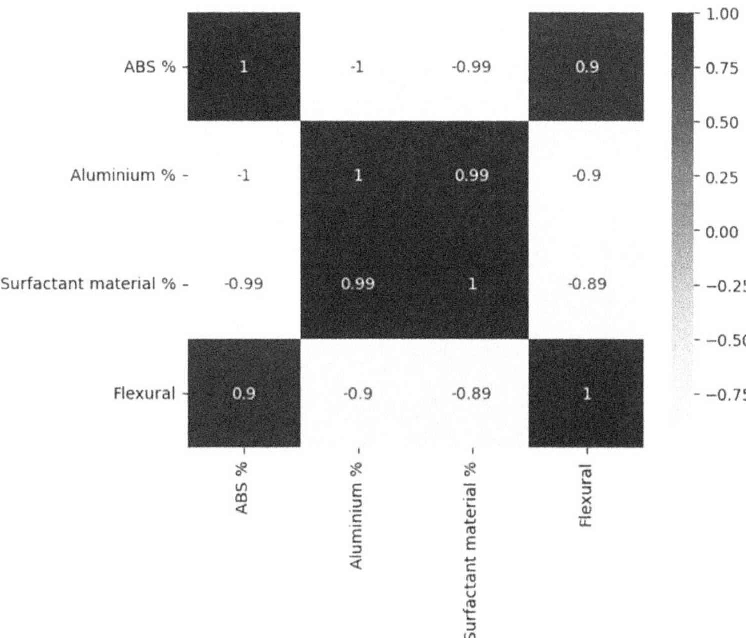

Figure 3(a): ABS-Al composition heatmap

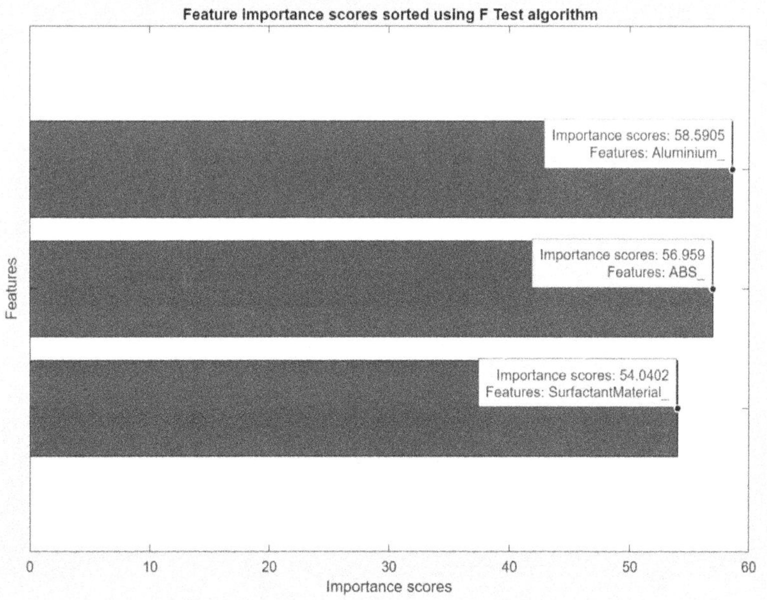

Figure 3(b): F-test for ABS-Al feature selection

Figure 4(a): ABS-Cu (Copper 60% + ABS 35%) after flexural test

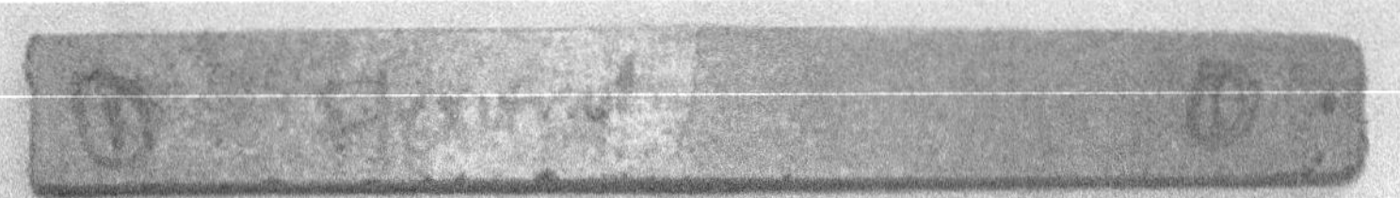

Figure 4(b): ABS-Cu (Copper 50% + ABS 44%) after flexural test

Figure 4(c): ABS-Cu (Copper 70% + ABS 23%) after flexural test

Figure 5(a): ABS-Al (Aluminum 30% + ABS 35%) after flexural test

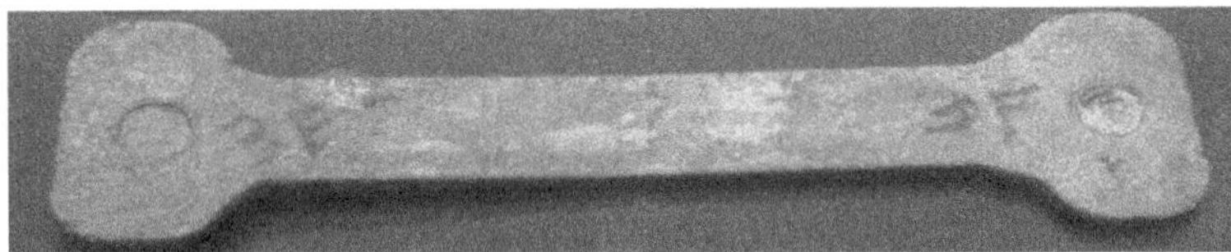

Figure 5(b): ABS-Al (Aluminum 50% + ABS 44%) after flexural test

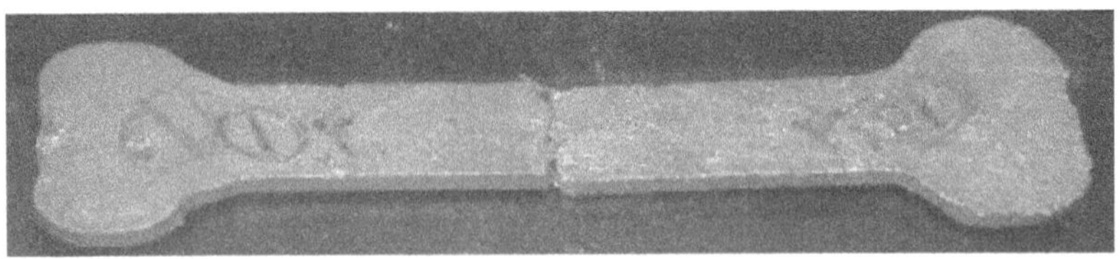

Figure 5(c): ABS-Al (Aluminum 70% + ABS 23%) after flexural test

3. False Negative in the bottom left quadrant 3.
4. Genuine Positivity (lower-right quadrant).

The precision, recall, f1-score, and support are shown in the classification report in Table 2(a) for the flexural strength prediction's precision and recall values are accurate because they are greater than 0.75, which indicates that the data was predicted with 75% accuracy or above. The AUC-ROC curve for ABS-Cu has a value of 0.7336, which is above 70%, showing that the classification model correctly predicted the findings for flexural strength.

Table 2a: Classification report for ABS-Cu

Classifier	Precision	Recall	F1- score	Support
0	1.00	0.62	0.77	8
1	0.80	1.00	0.89	12
Accuracy	-	-	0.95	20
Macro Avg	0.90	0.81	0.83	20
Weighted Avg	0.88	0.85	0.85	20

4.2 Machine Learning Classification: ABS-Al

Similar to this, the data for the flexural strength of ABS-Al composition was classified using k-nearest neighbours. The values above the average value of flexural strength which was 45.60 were considered as 1 and below the average values were considered as 0, For k=1, 2, 3, and 5, respectively, the classification accuracy for the k-nearest neighbour classification for the flexural strength of the ABS-Al composition was 85%, 95%, 95%, and 95%. The Euclidean distance between the nearest and average values of flexural strength are calculated using the k-nearest neighbour method. The classification report for the flexural strength of ABS-Al composition is presented in Table 2(b). It shows the precision, recall, f1-score, and support for the prediction accuracy of the classification model. The AUC-ROC curve for ABS-Cu shows a value of 0.8965, which is nearer to 1 as shown in Fig. 6(e), shows that the classification model accurately predicted the flexural strength values.

Table 2(b): Classification report for ABS-Al

Classifier	Precision	Recall	F1- score	Support
0	1.00	0.89	0.94	9
1	0.92	1.00	0.96	11
Accuracy	-	-	0.95	20
Macro Avg	0.96	0.94	0.95	20
Weighted Avg	0.95	0.95	0.95	20

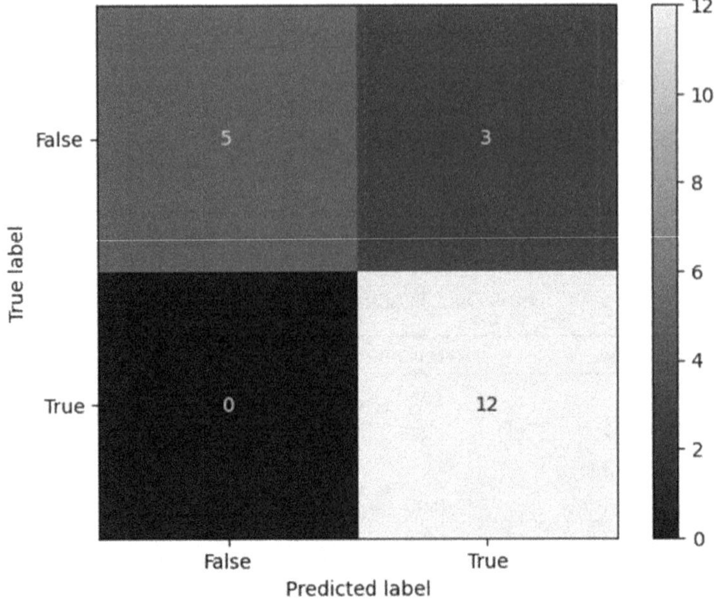

Figure 6(a): ABS-Cu confusion matrix for k=1

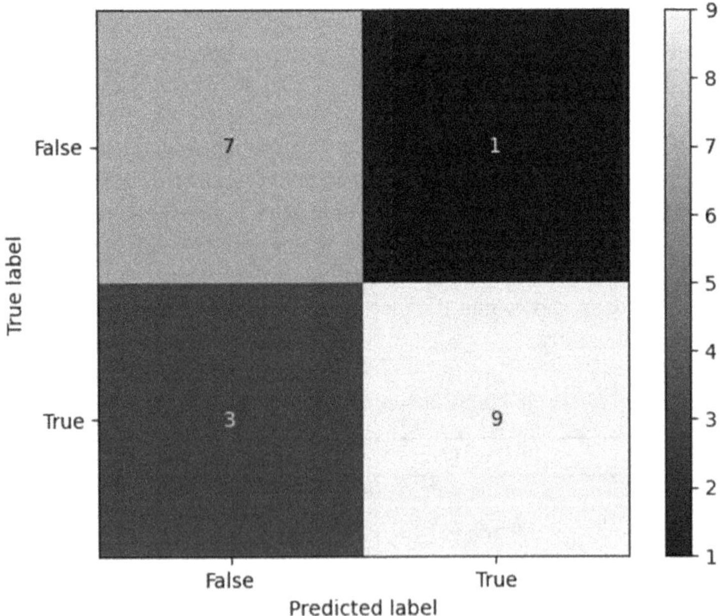

Figure 6(b): ABS-Cu confusion matrix for k=2

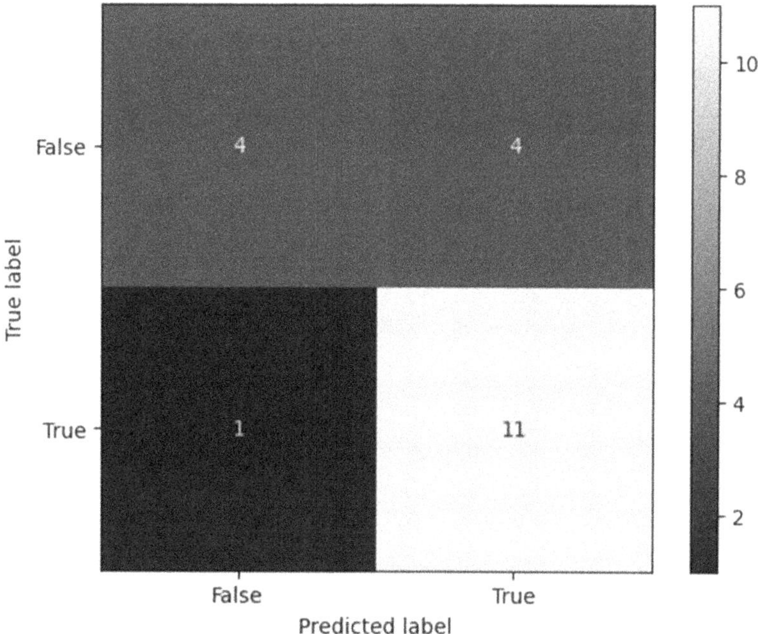

Figure 6(c): ABS-Cu confusion matrix for k=3

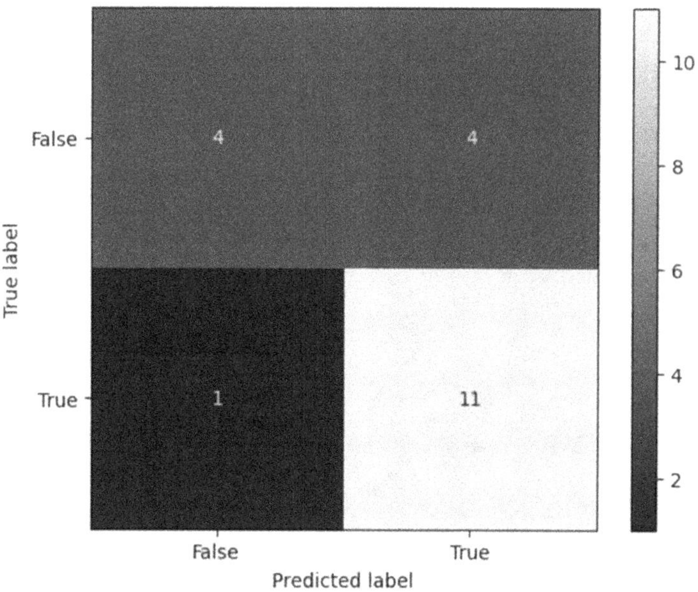

Figure 6(d): ABS-Cu confusion matrix for k=5

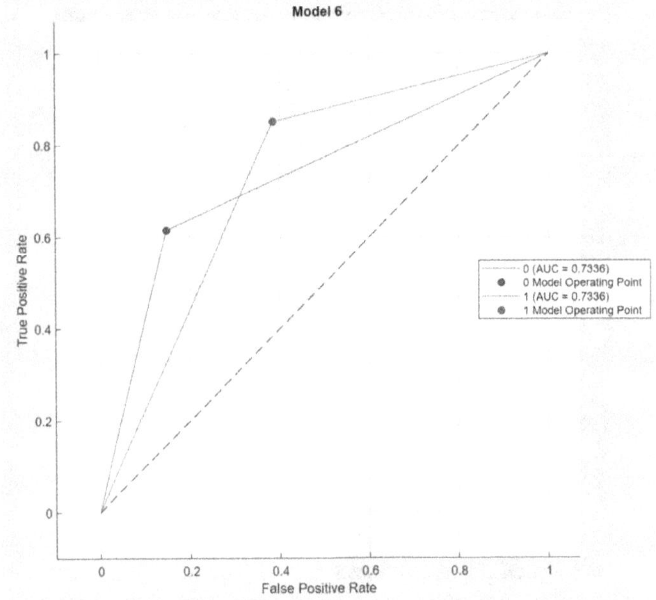

Figure 6(e): AUC-ROC curve for ABS-Cu

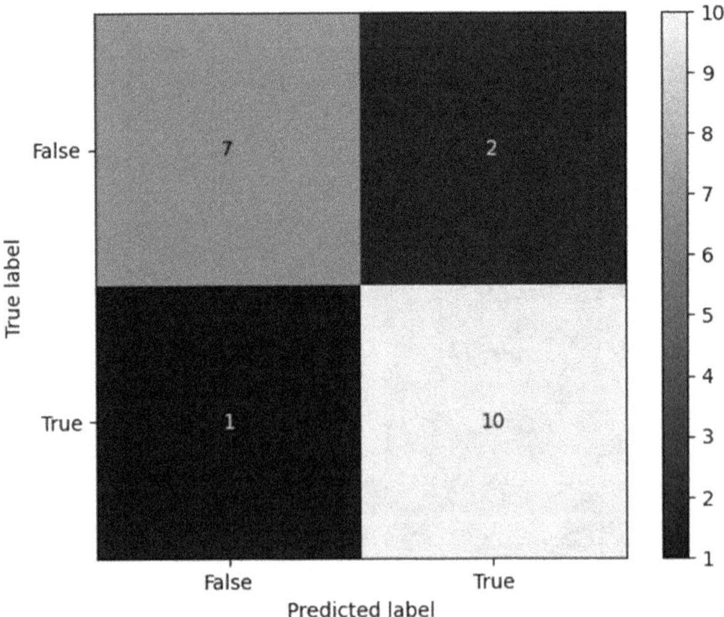

Figure 7(a): ABS-Al confusion matrix for k=1

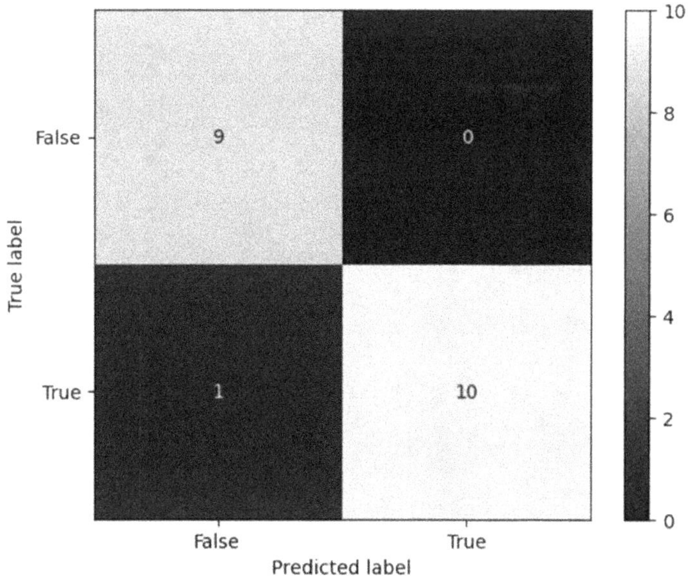

Figure 7(b): ABS-Al confusion matrix for k=2

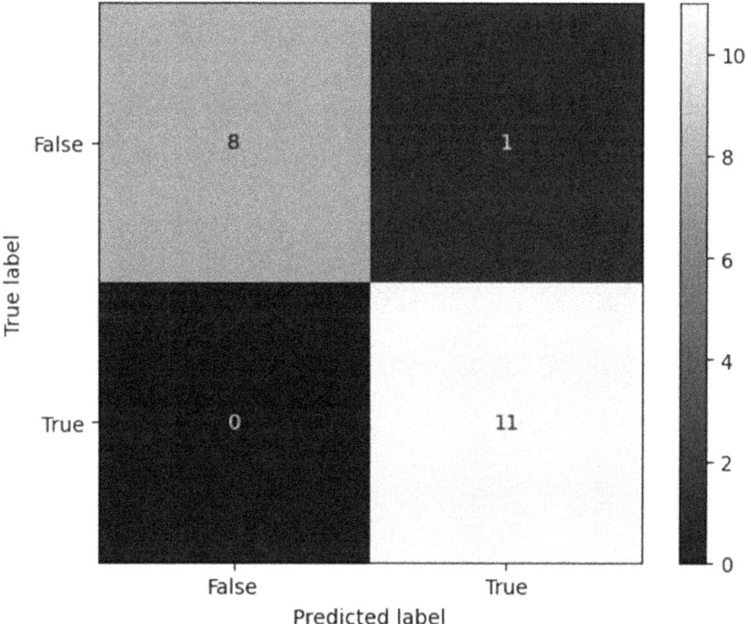

Figure 7(c): ABS-Al confusion matrix for k=3

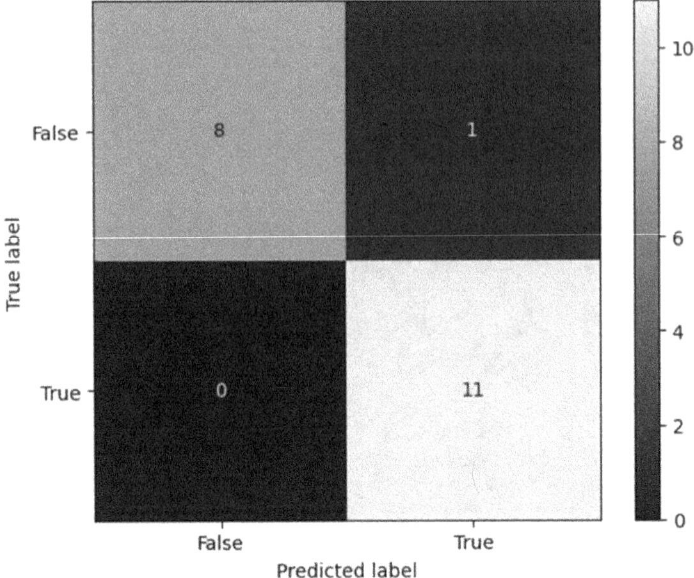

Figure 7(d): ABS-Al confusion matrix for k=5

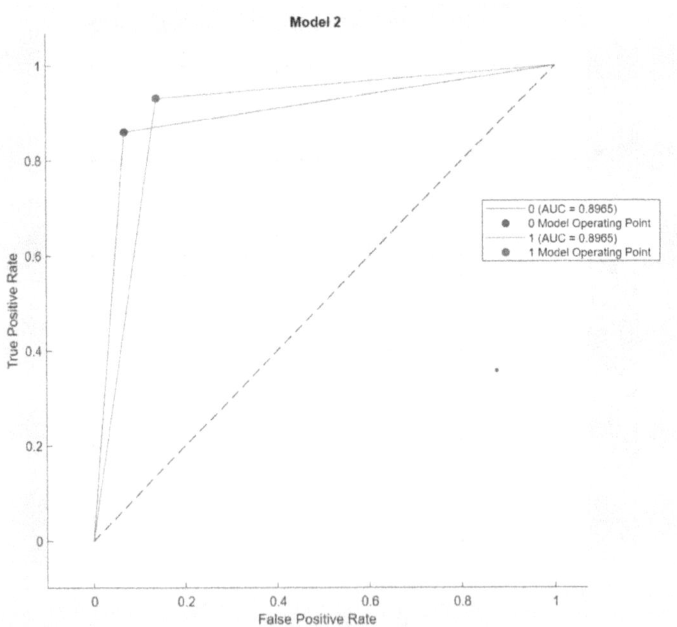

Figure 7(e): AUC-ROC curve for ABS-Al

5. Conclusion

This study's main goal was to determine how changes in the percentage compositions of ABS, copper, and aluminum in ABS-Cu and ABS-Al composite materials affected the materials' ability to flex. The percentage impacts of the surfactant, copper, and ABS in ABS-Cu were about equal to 10.8%, 4.55, and 3.5%, respectively, whereas the percentage effects of the surfactant, aluminum, and ABS were approximately equal to 58%. By showing confusion matrices for true positives, true negatives, false positives, and false negatives with values close to 100%, flexural strength was predicted using machine learning classification. For k=1, 2, 3, and 5, the prediction accuracy values for ABS-Cu were 85%, 80%, 75%, and 75%, respectively, and for ABS-Al composition, they were 85%, 95%, 95%, and 95%, respectively. These figures demonstrate that the prediction was accurate. The precision, recall, and accuracy values in the classification report are also closer to 1, demonstrating the high accuracy of the flexural strength prediction. Instead, a comprehensive analysis of the mechanical properties and strength of reinforced carbon fibres will be the focus of this work.

References

1. Karimah, A., Ridho, M.R., Munawar, S.S., Amin, I., Damayanti, Y., Lubis, R. et al. 2021. A comprehensive review on natural fibers: Technological and socio-economical aspects. *Polymers*, 13: 4280.
2. Joshi, S.V., Drzal, L.T., Mohanty, A.K. and Arora, S. 2004. Are natural fiber composites environmentally superior to glass fiber reinforced composites? *Compos. Part A Appl. Sci. Manuf.*, 35: 371–376.
3. Ravikumar, P., Rajeshkumar, G., Manimegalai, P., Sumesh, K.R., Sanjay, M.R. and Siengchin, S. 2022. Delamination and surface roughness analysis of jute/polyester composites using response surface methodology: Consequence of sodium bicarbonate treatment. *J. Ind. Text.*, 51: 360S–377S.
4. Sathish, T., Palani, K., Natrayan, L., Merneedi, A., De Poures, M.V. and Singaravelu, D.K. 2021. Synthesis and characterization of polypropylene/ramie fiber with hemp fiber and coir fiber natural biopolymer composite for biomedical application. *Int. J. Polym. Sci.*, 2462873.
5. Fortea-Verdejo, M., Bumbaris, E., Burgstaller, C., Bismarck, A. and Lee, K.Y. 2017. Plant fibre-reinforced polymers: Where do we stand in terms of tensile properties? *Int. Mater. Rev.*, 62: 441–464.
6. Kim, T., Tran, T.H., Hwang, S.Y., Park, J., Oh, D.X. and Kim, B.S. 2019. Crab-on-a-tree: All biorenewable, optical and radio frequency transparent barrier nanocoating for food packaging. *ACS Nano*, 13: 3796–3805.
7. Nugroho, G. and Budiyantoro, C. 2022. Optimization of fiber factors on flexural properties for carbon fiber reinforced polypropylene. *J. Compos. Sci.*, 6: 160.
8. Hashemi, S. 2008. Hybridisation effect on flexural properties of single- and double-gated injection moulded acrylonitrile butadiene styrene (ABS) filled with short glass fibres and glass beads particles. *J. Mater. Sci.*, 43: 4811–4819.

9. Serra-Parareda, F., Vilaseca, F., Aguado, R., Espinach, F.X., Tarrés, Q. and Delgado-Aguilar, M. 2021. Effective Young's modulus estimation of natural fibers through micromechanical models: The case of henequen fibers reinforced–PP composites. *Polymers*, 13: 3947.
10. Tarrés, Q., Oliver-Ortega, H., Espinach, F.X., Mutjé, P., Delgado-Aguilar, M. and Méndez, J.A. 2019. Determination of mean intrinsic flexural strength and coupling factor of natural fiber reinforcement in polylactic acid biocomposites. *Polymers*, 11: 1736.
11. Yamaguchi, A., Urushisaki, M., Uematsu, H., Sakaguchi, T. and Hashimoto, T. 2022. Effects of different types of maleic anhydride-modified polypropylene on the interfacial shear strengths of carbon fiber-reinforced polypropylene composites. *Polym. J.*, 38: 1–9.
12. Mirbagheri, J., Tajvidi, M., Hermanson, J.C. and Ghasemi, I. 2007. Tensile properties of wood flour/kenaf fiber polypropylene hybrid composites. *J. Appl. Polym. Sci.*, 105: 3054–3059.
13. Aisyah, H.A., Paridah, M.T., Sapuan, S.M., Ilyas, R.A., Khalina, A., Nurazzi, N.M. et al. 2021. A comprehensive review on advanced sustainable woven natural fibre polymer composites. *Polymers*, 13: 471.
14. Sanadi, A.R., Young, R.A., Clemons, C. and Rowell, R.M. 1994. Recycled newspaper fibers as reinforcing fillers in thermoplastics: Part I – Analysis of tensile and impact properties in polypropylene. *J. Reinf. Plast. Compos.*, 13: 54–67.
15. Lopez, J.P., Girones, J., Mendez, J.A., El Mansouri, N.E., Llop, M. and Mutjé, P. 2012. Stone-ground wood pulp-reinforced polypropylene composites: Water uptake and thermal properties. *BioResources*, 7: 5478–5487.

CHAPTER

10

Supervised Machine Learning Based Classification of Surface Roughness of Fused Deposition Modeling 3D Printed Samples

Rithvik Nair[1], Shreya Joshi[1], Kaustubh Dwivedi1, Bhargav Gadhiya[2], Mandar S. Sapre[1*], Ashwini V. Jatti[3]

[1] Symbiosis Institute of Technology, Symbiosis International (Deemed University), Pune, Maharashtra, India

[2] ISB&M College of Engineering, Savitribai Phule Pune University, Pune, Maharashtra, India

[3] D.Y. Patil Institute of Technology, Savitribai Phule Pune University, Pune, Maharashtra, India

1. Introduction

Machine learning is a subset of artificial intelligence that involves the use of algorithms and statistical models to analyze data and make predictions or decisions. In the context of Fused Deposition Modelling (FDM), machine learning can be used in several ways to improve the quality, efficiency, and reliability of the process.

Fused Deposition Modelling (FDM) printing is one of the most popular and widely used 3D printing technologies due to its affordability, ease of use, and versatility. FDM printers are available at various price points, making them accessible to hobbyists, small businesses, and large manufacturers alike. It is a type of additive manufacturing technology used in 3D printing. It works by melting a thermoplastic filament and extruding it layer by layer to create a 3D object.

Various researchers have worked on similar grounds and have found many important relations between printing parameters and the physical properties of

*Corresponding author: mandar.sapre@sitpune.edu.in

printed products. The parameters commonly used were bed temperature, nozzle temperature, printing speed, infill density, layer height, and other printing parameters [1-3]. These print parameters are the ones governing the physical state of the product printed. The overall print quality, deviation (error) from the original dimensions, the surface finish, the overall smoothness of the product etc. directly or indirectly depends on these printing parameters [4-6].

Machine Learning models have been used recently in the field of additive manufacturing to train the models to predict output for optimization, though this whole idea is still growing. Classification and regression models have been used in training and predicted output parameters, and this data has been used for optimization [7, 8].

Classification algorithms can be used to predict print quality based on input parameters such as layer height, nozzle temperature, print speed, and material type. Regression algorithms can be used to predict the mechanical properties of FDM parts based on these same input parameters [9]. Deep learning techniques, such as convolutional neural networks (CNNs) and long-, short-term memory (LSTM) networks, can be used to extract features from the input data and improve the accuracy of these predictions [8, 10].

Improving the quality of printed objects and streamlining the FDM manufacturing process are potential benefits of machine learning. Machine learning can optimize print settings and reduce print time, resulting in more efficient and cost-effective manufacturing processes. This study specifically looks at the impact of FDM printing parameters on surface roughness. Several process parameters were chosen based on recent studies that influence surface roughness and three classification models were used to classify and test the dependence of the printing parameters on surface roughness. These predictions can assist in further optimization of print parameters to achieve a smoother finish.

The paper by Chand et al. [4] investigates the effect of FDM printing parameters on surface roughness. The study involved observing and noting down a series of process parameters that directly influence the surface roughness. To classify and test the dependence of printing parameters on surface roughness, the authors used three different classification models: the k-Nearest Neighbours algorithm (kNN), the Naïve-Bayes algorithm, and Logistic regression. The machine learning models' predictions can significantly help in further optimization of the print parameters to obtain a smoother finish.

Several other studies have also examined the quality of 3D-printed parts. For instance, Hanon et al. [11] investigated the effects of various parameters, such as layer thickness, printing speed, and extruder temperature on the accuracy of 3D-printed parts. Meanwhile, Mendricky and Fris [1] experimentally determined optimal process parameters to improve the quality of 3D-printed parts. In another study, Selvam et al. [12] optimized the surface roughness and printing time of FFF 3D printed parts using a multi-objective optimization approach. Finally, Gomez-Gras et al. [2] compared the accuracy and surface quality of holes produced using FDM 3D printing technology with those produced by machining in PEI Ultem 9085 specimens. The findings of these studies can be useful in enhancing the performance of FDM/FFF technology and improving the quality of 3D-printed parts.

This paper is organized in such a way that it allows the reader to understand and quickly come to terms with the dependency of surface roughness quickly and systematically. The introduction allows the reader to get a brief overview of the study, and this is followed by a literature review which briefs the recent study in the domain. Then the methodology section explains the overall experimental setup and the applied print parameters. The overview of each of the classification models and other important parameters used to test the credibility of the results are also explained in this section. In the results and discussions section, the results of the classification are analyzed and discussed and the insights gained are properly summarized. Overall, the paper structure ensures that the reader can easily follow and understand how different printing parameters directly affect the surface roughness of the product.

2. Methodology

2.1 Workflow

To have a consistent model, the geometry of the ASTM E8 standard was selected as the benchmark geometry. The dimensions were uniformly scaled down by 50% to reduce the print size and thickness was set to 5 mm, the amount of required material and time. Thirty different trial conditions were constructed using Response Surface Methodology (RSM) Design of the Experiment method and on three levels of each input parameter. The CAD model was sliced according to these trial conditions and G-code was generated using UltimakerCura software.

The experimental investigation was conducted using a Creality 3D FDM printer. Each print received a unique set of settings. Thirty prints with various layer heights, infill densities, infill patterns, bed temperatures, and nozzle temperatures were produced. Then, a datasheet was created to organise the input parameters. Additionally, the differences between each model's length, breadth, and thickness of the original CAD file that we created were calculated with the aid of a digital vernier calliper. Figure 1 illustrates the stepwise sequence for entire work. Figure 2a depicts the CAD drawing of specimen and Figure 2b shows the actual 3D printed specimens. Figure 2c shows the 3D printer used in present study.

These print parameters are then randomly split into two sets with twenty-four sets and six sets (and 4:1 split) to train classification models and evaluate them, respectively. The first set (training set) is fed to the machine learning application, which will return the correlation matrix, which will return the dependency of the input parameters on each other as well as the fed output parameters.

Following this, classification models were applied to the training and testing data respectively, based on which the performance evaluation was carried

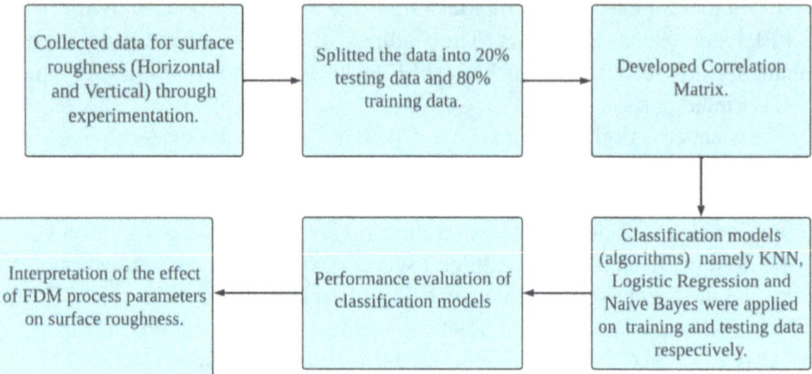

Figure 1: Workflow diagram

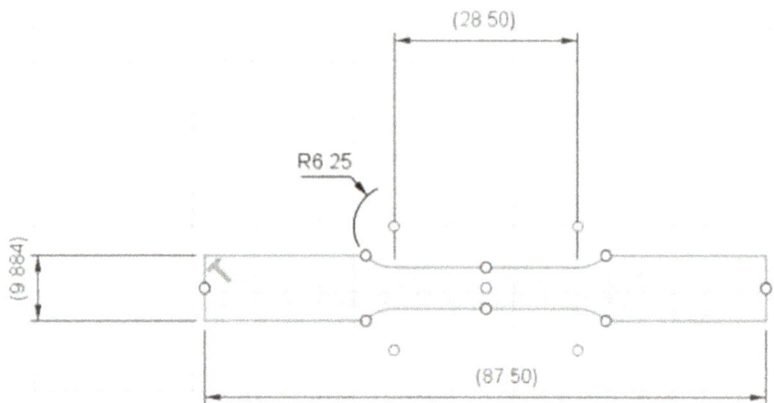

Figure2(a): CAD drawing of the model with dimensions

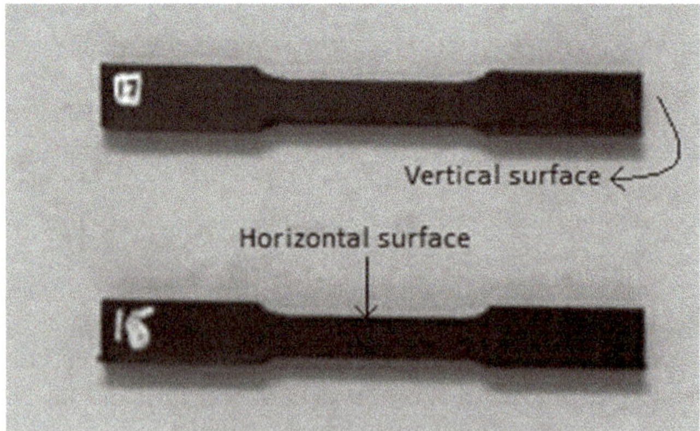

Figure 2(b): Final printed model

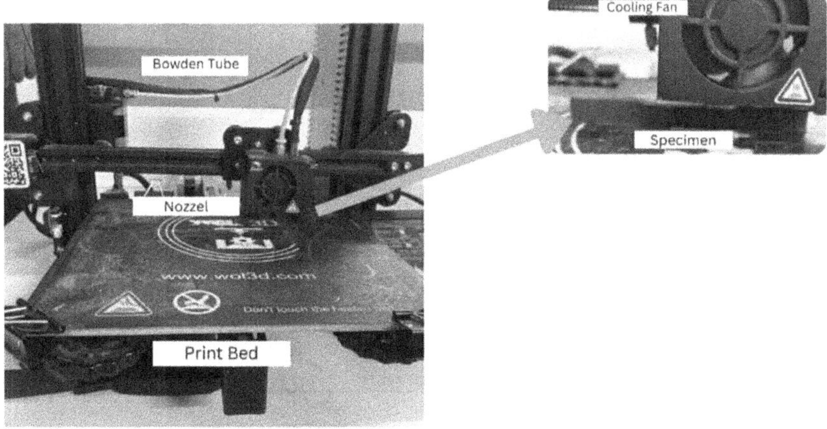

Figure 2(c): Creality-3D FDM printer

out after the results of these models was obtained. In this study, classification algorithms applied were the K-Nearest Neighbour (kNN Algorithm), the Naïve Bayes Algorithm and Logistic Regression. For this the average surface roughness of the 30 prints was computed and classed into two (0 and 1) based on whether they were higher or lower than the mean. Overall, the study intended to find out the effect of FDM print parameters on the surface roughness of the final printed product using the above classification models, and the resulting findings can be used for the optimization of the same print parameters for a better-quality print.

2.2 Background of the Classification Algorithms Applied

K Nearest Neighbour

The k-Nearest Neighbours (kNN) algorithm is a widely used classification algorithm in machine learning. The kNN algorithm searches the k-nearest neighbours and works by converging to a new data point in the feature space. These k-nearest neighbours are found by a distance metric, like Euclidean distance or Manhattan distance.

$$\text{Minkowski distance} = \left(\sum_{i=1}^{n} \left|x_i - y_i\right|\right)^{\frac{1}{p}} \tag{1}$$

Minkowski distance is a mathematical concept used to measure the similarity or dissimilarity between two points in a multi-dimensional space. It is a generalization of several other distance metrics, including Euclidean distance and Manhattan distance, and it allows you to control the level of sensitivity to differences in various dimensions by using a parameter known as "p".

In equation 1, x_i and y_i denote the respective co-ordinates of the points X and Y in the ith dimension. The parameter p is a positive real number. When this

Table 1: Print parameters, calculated deviations, and measured surface roughness

Trail condition	Layer height (mm)	Wall thickness (mm)	Infill density (%)	Infill Pattern	Nozzle tem. (°C)	Bed tem. (°C)	Print speed (nim/sec)	Fan speed (%)	Length wise deviation	Breadth wise deviation	Height wise deviation	Surface roughness (vertical)	Surface roughness (horizontal)
1	0.1	1	50	Honeycomb	200	60	120	0	0.34	0.1539	0.025	0.26525	6.12275
2	0.1	4	40	Grid	205	65	120	25	0.13	0.0589	-0.075	0.314	6.35675
3	0.1	3	50	Honeycomb	210	70	120	50	0.28	0.0539	0.045	0.288	5.957
4	0.1	4	90	Grid	215	75	120	75	0.36	0.0839	0.045	0.22875	5.92025
5	0.1	1	30	Honeycomb	220	80	120	100	0.22	0.2039	-0.005	0.328	6.08775
6	0.15	3	80	Honeycomb	200	60	60	0	0.2	0.2977	0.035	0.322	6.0684
7	0.15	4	50	Grid	205	65	60	25	0.26	0.1389	0.005	0.44125	9.27525
8	0.15	10	30	Honeycomb	210	70	60	50	0.18	0.0964	0.025	0.5291	7.479
9	0.15	6	40	Grid	215	75	60	75	0.36	0.2839	0.065	0.30425	7.557
10	0.15	1	10	Honeycomb	220	80	60	100	0.15	0.1839	0.035	0.4355	8.48675
11	0.2	5	60	Honeycomb	200	60	40	0	0.12	0.4039	0.015	0.3685	8.4695
12	0.2	4	20	Grid	205	65	40	25	0.14	-0.0411	-0.025	0.3665	8.8785
13	0.2	5	60	Honeycomb	210	70	40	50	0.1	-0.0561	0.195	0.091	9.415
14	0.2	7	40	Grid	215	75	40	75	0.1	0.0614	0.015	0.52825	9.71375
15	0.2	3	60	Honeycomb	220	80	40	100	0.11	0.7789	0.015	0.25975	10.59625
16	0.1	1	50	Triangles	200	60	120	0	0.06	0.2614	0.025	0.221	6.04925
17	0.1	4	40	Cubic	205	65	120	25	0.27	0.1364	0.135	0.425	9.262

(Contd.)

Table 1: (*Contd.*)

18	0.1	3	50	Triangles	210	70	120	50	0.21	0.0339	0.025	0.1955	6.127
19	0.1	4	90	Cubic	215	75	120	75	0.46	0.2014	0.075	0.372	5.99675
20	0.1	1	30	Triangles	220	80	120	100	0.21	0.1814	0.045	0.30275	6.1485
21	0.15	3	80	Triangles	200	60	60	0	0.25	0.2039	0.035	0.306	8.2585
22	0.15	4	50	Cubic	205	65	60	25	0.2	0.2364	-0.005	0.2635	8.347
23	0.15	10	30	Triangles	210	70	60	50	0.25	0.2314	-0.085	0.41875	8.2385
24	0.15	6	40	Cubic	215	75	60	75	0.38	0.2364	0.065	0.28625	8.23125
25	0.15	1	10	Triangles	220	80	60	100	0.19	0.2964	0.035	0.16775	8.35125
26	0.2	5	60	Triangles	200	60	40	0	0.17	0.2039	-0.015	0.343	9.072
27	0.2	4	20	Cubic	205	65	40	25	0.14	0.2789	0.055	0.25525	9.23825
28	0.2	5	60	Triangles	210	70	40	50	0.27	0.4239	0.015	0.26325	9.18225
29	0.2	7	40	Cubic	215	75	40	75	0.21	0.3464	0.065	0.25175	9.299
30	0.2	3	60	Triangles	220	80	40	100	0.08	0.0539	0.015	0.19075	9.382

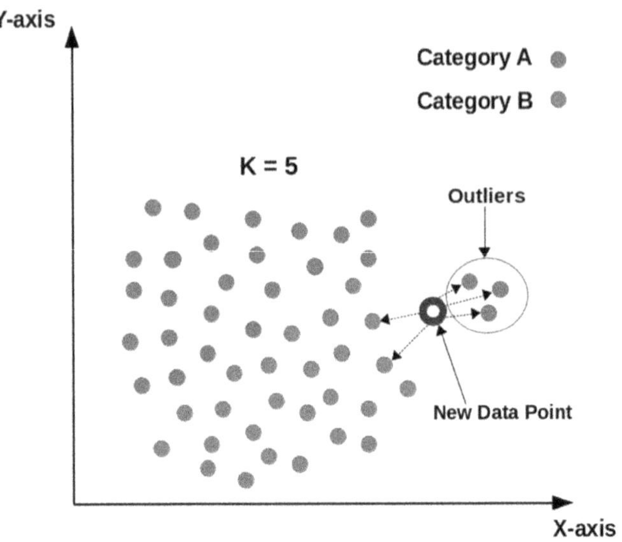

Figure 3: Pictorial representation of kNN algorithm

is equal to 1, it computes the Manhattan distance, while at 2, it computes the Euclidean distance.

Naïve Bayes

The Naive Bayes classifier is a probability-based classification algorithm used in machine learning. The classifier uses the Bayes' Theorem, which computes the probability of an event based on earlier knowledge of some conditions directly or indirectly related to the event.

$$P(A|B) = \frac{P(B \mid A)P(A)}{P(B)} \tag{2}$$

In the formula as given in equation 2:

- $P(A|B)$ is the probability of class A given the observed features B. This is the quantity we want to compute, the probability that a data point with features A belongs to class B.
- $P(B|A)$ is the probability of observing the features B given that the data point belongs to class A. This is called the likelihood.
- $P(A)$ is the prior probability of class A. It represents our initial belief or probability that a data point belongs to class C before considering any features.
- $P(B)$ is the probability of observing the features B across all classes. It acts as a normalizing constant to ensure that the probabilities sum up to 1.

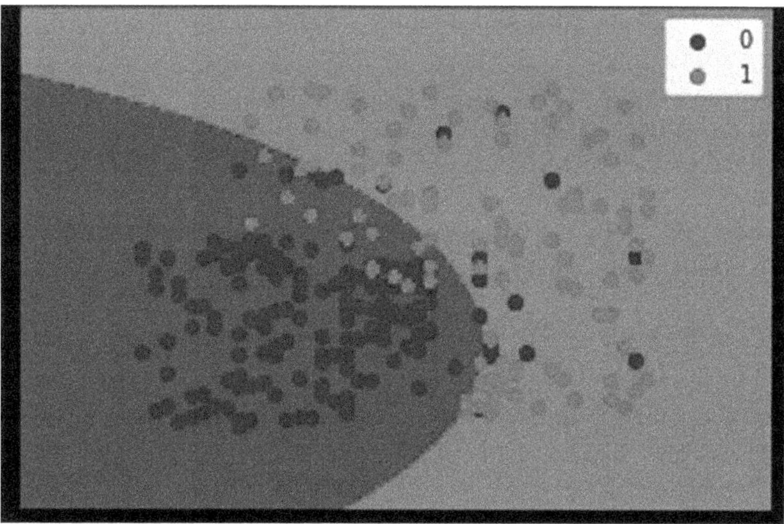

Figure 4: Pictorial representation of the Naïve Bayes classifier

Logistic Regression

Logistic regression is a classification algorithm used in machine learning. It models the relation between a set of input values and a final binary value. The logistic regression algorithm works by learning the coefficients of the input features that maximize the likelihood of the observed training data.

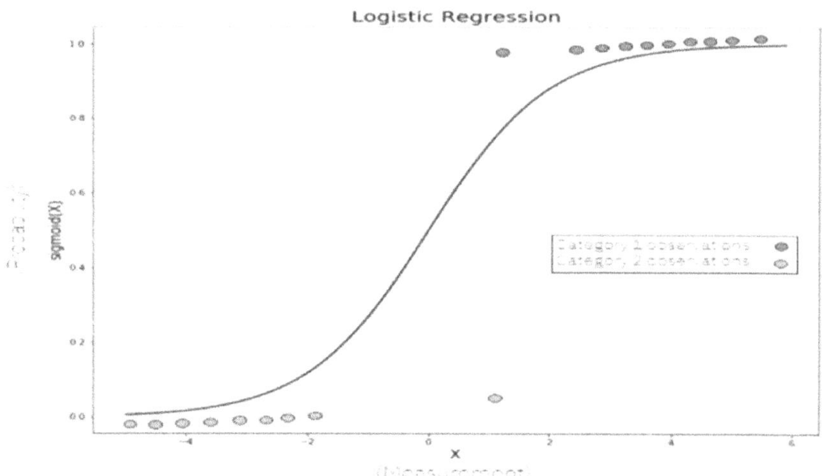

Figure 5: Pictorial representation of the logistic regression classifier

2.3 Some Important Supervised Learning Terms

Precision: Precision is a commonly used evaluation metric in machine learning, particularly for classification tasks. Precision measures the number of positive cases that were predicted to be positive. So, it measures the ratio between true positives (TP) to the actual positive predictions. The formula is given in equation 3.

Precision for is defined as the ratio of number of true positives to the number of predicted positives.

$$\text{Precision} = \frac{\text{True Positive}}{\text{False Positive} + \text{True Positive}} \tag{3}$$

Recall (Sensitivity): Recall is a commonly used evaluation metric in machine learning, particularly for classification tasks. Recall measures the number of positive cases that were correctly identified by the model. So, it measures the ratio between true positives (TP) to the actual positive cases. The formula is given in equation 4.

Recall is defined as the ratio of number of true positives to the total number of actual positives.

$$\text{Recall} = \frac{\text{True Positive}}{\text{False Negative} + \text{True Positive}} \tag{4}$$

F1 Score: It gives a relative idea of Precision and Recall. It is maximum when Precision is equal to Recall. The formula is given in equation 5.

F1 Score is the harmonic mean of precision and recall.

$$\text{F1} = 2.\frac{\text{Precision} \times \text{Recall}}{\text{Precision} + \text{Recall}} \tag{5}$$

AUC-ROC: AUC-ROC stands for "Area Under the Receiver Operating Characteristic Curve". It is a metric commonly used in machine learning to evaluate the performance of a binary classification model, which predicts the probability of an input belonging to one of two classes. The ROC curve is a graphical representation of the performance of a binary classifier as its discrimination threshold is varied. It plots the true positive rate (sensitivity) against the false positive rate (1-specificity) at different threshold values. The AUC-ROC is the area under this curve, which ranges from 0 to 1, with a higher value indicating a better classifier performance. In essence, the AUC-ROC measures the ability of the model to distinguish between positive and negative classes, regardless of the chosen classification threshold. A perfect classifier has an AUC-ROC score of 1, while a random classifier has an AUC-ROC score of 0.5.

Specificity: Specificity is another commonly used evaluation metric in machine learning, particularly for binary classification tasks. It measures the proportion of

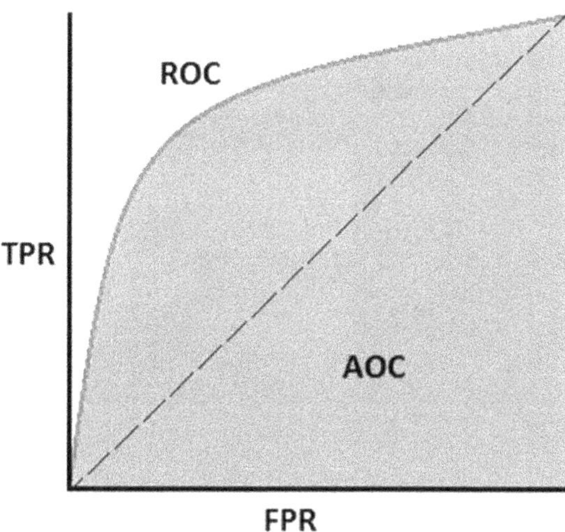

Figure 6: Receiver- Operator Characteristic (ROC)

true negative (TN) predictions among all the actual negative cases in the dataset. The formula is given in equation 6.

Specificity is defined as how many of the negative cases were correctly identified by the model.

$$\text{Specificity} = \frac{\text{True Negative}}{\text{True Negative} + \text{False Positive}} \tag{6}$$

3. Results and Discussions

3.1 Effect of FDM Process Parameters on Surface Roughness

Using Machine Learning models, we can find the correlation matrix (also called a heatmap) which will allow us to read into the process parameters and find the dependency of the surface roughness on them. This is a key step to understand what process parameters can be changed and by what amount they must be changed to get a subsequent and desirable change in surface roughness, as achieving a smooth surface finish in FDM requires finding the right balance between these process parameters.

Figure 7 shows the correlation matrix of vertical surface roughness. It is observed that in this case, it depends more on wall thickness. All the differences are seen in the first two parameters, while the dependence on the rest of the parameters remains the same.

Figure 8 shows the correlation matrix of horizontal surface roughness. As seen, layer height dominates the dependence of horizontal surface roughness.

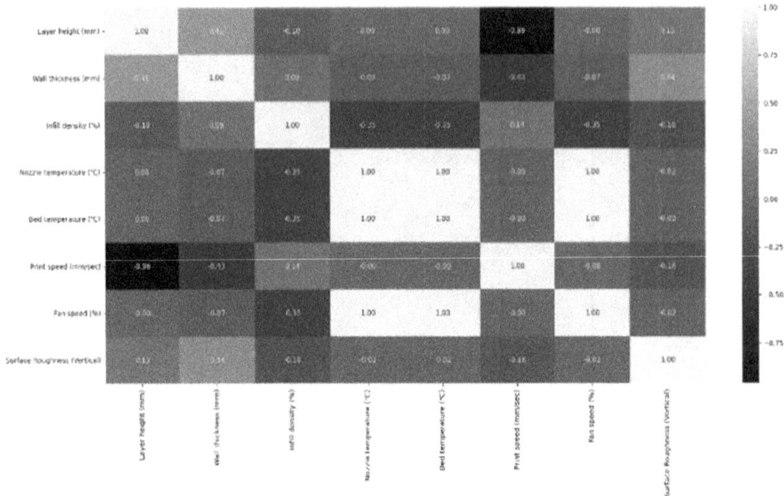

Figure 7: Correlation matrix for vertical surface roughness

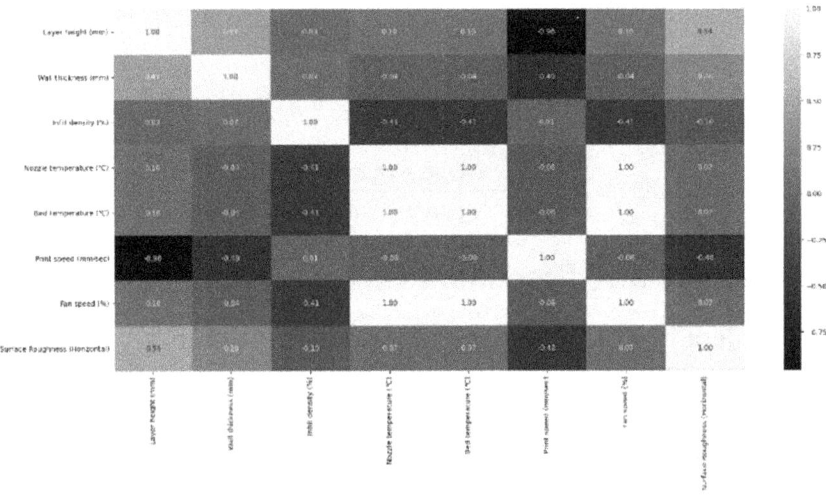

Figure 8: Correlation matrix for horizontal surface roughness

This means that a change in the layer height will bring the most deviation in the value of the vertical surface roughness.

3.2 kNN Results

Vertical Surface Roughness

For class 0, we get a precision of 0.67, which means 67% of the positive predictions were true positives. A recall of (0.61) shows that the model fairly predicted all the true positives. Meanwhile, report class 1 has precision (0.58), meaning that when it predicts positive cases, it is moderately correct. However, it has relatively better recall (0.636), indicating that it predicts a fair portion of actual positive cases. The F1 score (0.61) strikes a balance between precision and recall, suggesting a reasonably good overall performance. The specificity of 0.61 means that the model was able to predict all the true negatives fairly. Finally, the AUC ROC score of 62.59% suggests that the model has an average predictive power, with a 62.59% chance of correctly identifying any randomly selected positive instance. Overall, the results suggest that this model has a superior performance in identifying true positives in comparison to its ability to correctly identify negative instances, the latter of which needs improvement. Figure 9(a) and (b) depicts the Confusion matrix and ROC for kNN train model for vertical surface roughness, respectively. Figure 10(a) and (b) depicts the Confusion matrix and ROC for kNN test model for vertical surface roughness, respectively

Based on the values computed for class 0, we found that the model has a high recall (1), which means the model correctly identifies all the positive cases in the test set. However, the precision is low (0.33), which suggests that most of the positive predictions made by the model are false positives. For class 1, the model has perfect precision (1.0), meaning that when it predicts positive cases, it is always correct. However, it has a recall of 0, indicating that it fails to identify any of the actual positive cases. The F1 score is also 0.67, reflecting good overall performance. This suggests that the model's predictions are not capturing any true positives, which is a significant issue in classification tasks. The specificity is high (1), which means that the model can correctly identify all the negative cases in the dataset. The AUC ROC is 75%, which suggests that the model can properly distinguish the positive and negative cases effectively.

Horizontal Surface Roughness

For class 0, the model has a high recall (0.86), which means that it is correctly predicting the positive cases in the training set. The precision is also relatively high (0.81), indicating that the model is generating fewer false positives compared to true positives. For class 1, the model has a good precision (0.75), meaning that it correctly predicts a major number of positive cases. It also has a good recall (0.66), indicating that it captures over half of the actual positive cases. The F1 score (0.71) reflects a strong overall performance with a balanced trade-off between precision and recall. This suggests that your model is effective at both identifying positive cases and minimizing false positives, making it a solid choice for the task at hand. The specificity is high (0.86), which means that the model correctly identified majority of the negative cases in the dataset. Additionally, the AUC ROC is 76.67%, which shows that the model can distinguish between the positive and negative cases significantly well. Figure 11(a) and (b) depicts the Confusion matrix and ROC for kNN train model for horizontal surface roughness, respectively.

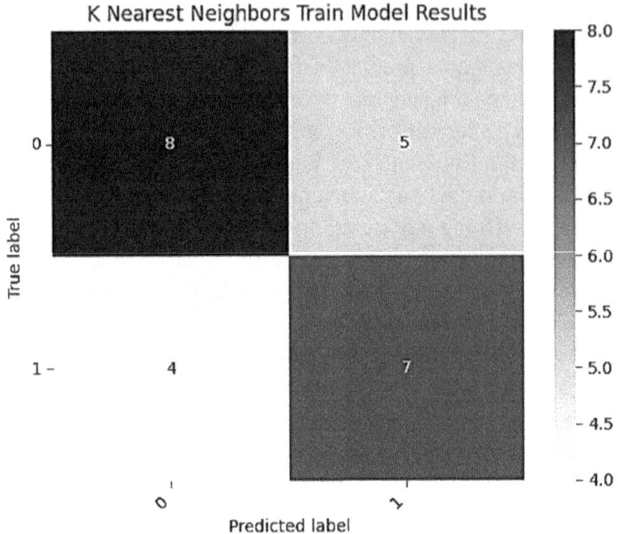

Report Class 0:
Precision = 0.67
Recall = 0.6153846153846154

Report Class 1:
Precision = 0.58
Recall = 0.6363636363636364

F1_Score = 0.61
Specificity = 0.6153846153846154
AUC ROC = 62.59%

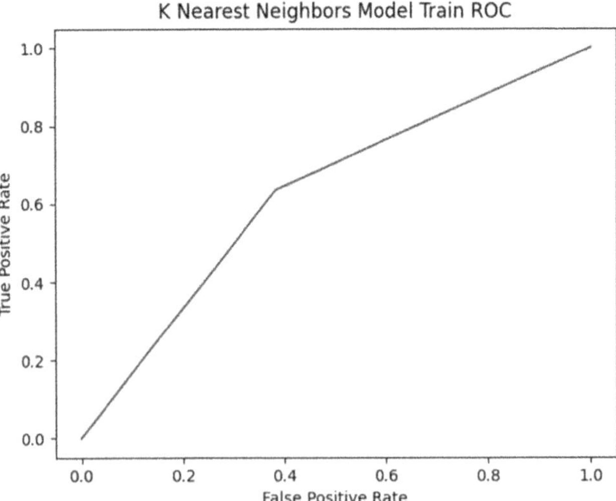

Figure 9(a) and (b): Confusion matrix and ROC for kNN train model for vertical surface roughness

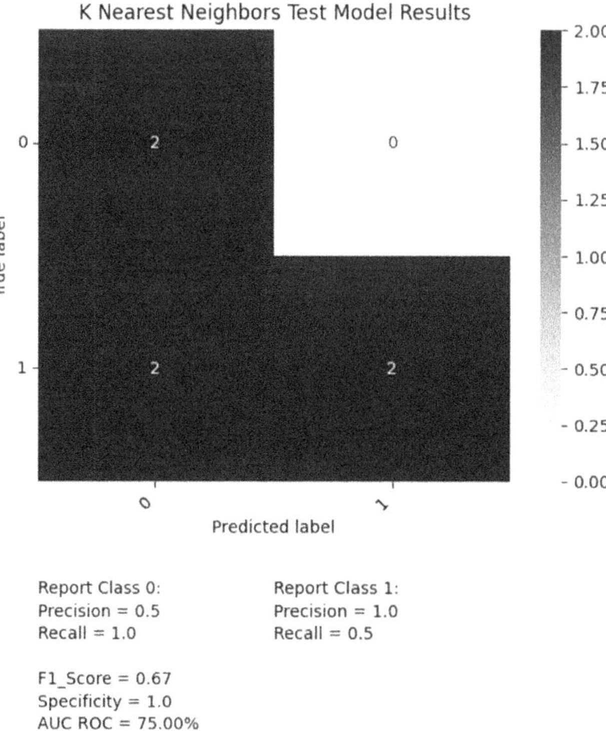

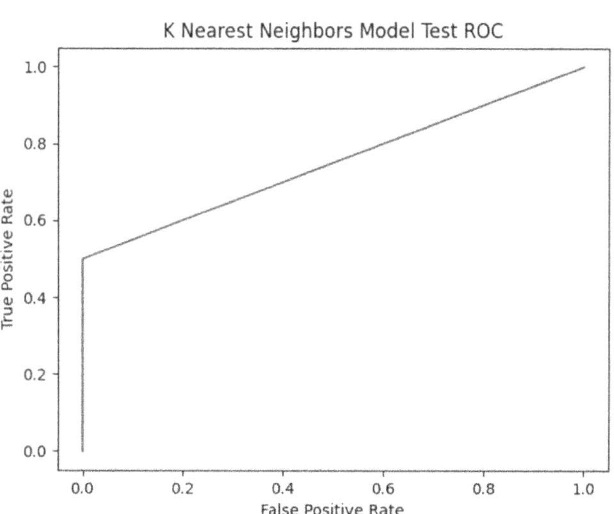

Figure 10(a) and (b): Confusion matrix and ROC for kNN test model for vertical surface roughness

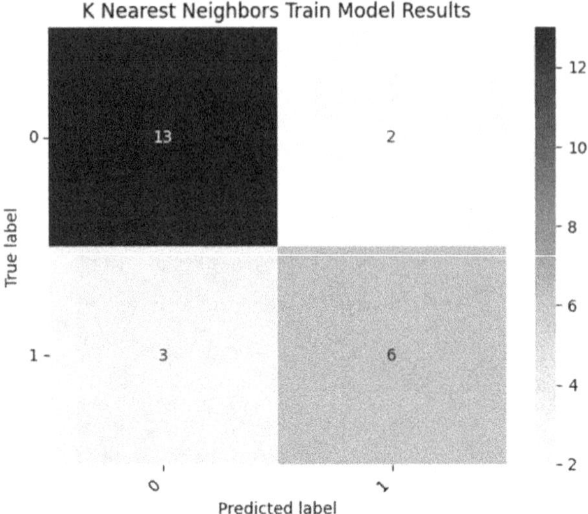

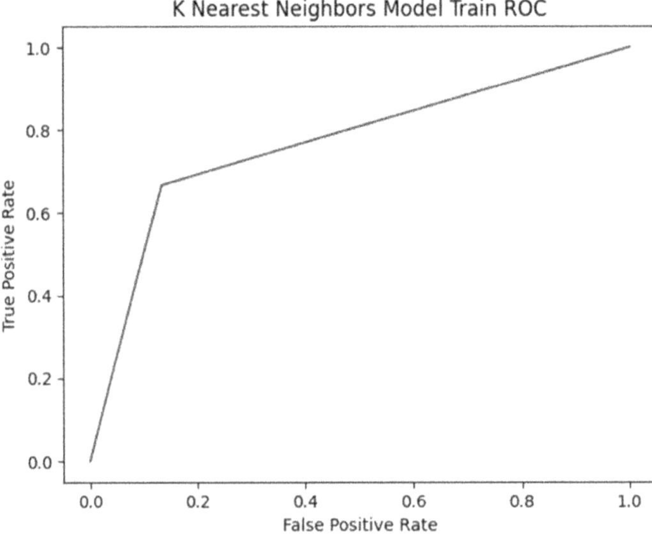

Figure 11(a) and (b): Confusion matrix and ROC for kNN train model for horizontal surface roughness

For class 0, the model has a high recall (1), which means that it is properly identifying the positive cases in the testing set. However, the precision is average (0.5), which indicates that the model is generatingmany false positives as compared to true positives. The F1 score is 0.67, indicating an average overall performance of the model. For class 1, the model has perfect precision (1.0), meaning that when it predicts positive cases, it is always correct. However, it has a recall of 0.5, indicating that it identifies about half of the actual positive cases. While the precision is high, the relatively low recall suggests that the model may miss a significant portion of positive cases. Depending on the specific application, this trade-off may or may not be acceptable. The F1 score is 0.67, indicating an average overall performance of the model. The specificity is high (1), which means that the model correctly identified all the negative cases in the dataset. However, the AUC ROC is 75%, which implies that the model can distinguish between the positive and negative cases fairly. Figure 12(a) and (b) depicts the Confusion matrix and ROC for kNN test model for horizontal surface roughness, respectively

3.3 Naïve-Bayes Results

Vertical Surface Roughness

For class 0, the model has a moderate precision (0.64), indicating that the model is generating a moderate number of false positives compared to true positives. The recall is also moderate (0.54), implying that the model is appropriately identifying some of the positive cases in the training set but not all of them. For class 1, the model has moderate precision (0.54), meaning that when it predicts positive cases, it is correct about 54% of the time. It also has a decent recall (0.64), indicating that it captures approximately 64% of the actual positive cases. The F1 score (0.58) represents a reasonable balance between precision and recall, indicating an overall acceptable performance. While there is room for improvement, especially in precision, the model is reasonably effective at identifying positive cases while maintaining a certain level of precision. The specificity is low (0.53), which meaning that the model is not able to correctly identify all the negative cases in the training set. Additionally, the AUC ROC is 58.74%, which suggests that the model cannot distinguish between the positive and negative cases effectively. Figure 13(a) and (b) depicts the Confusion matrix and ROC for Naïve-Bayes train model for vertical surface roughness, respectively.

For class 0, the model has a high recall (1), which means that it is correctly identifying all the positive cases in the test set. However, the precision is low (0.4), indicating that the model is generating a significant number of false positives compared to true positives. For class 1, the model has perfect precision (1.0), meaning that when it predicts positive cases, it is always correct. However, it has a low recall (0.25), indicating that it captures only a quarter of the actual positive cases. The F1 score (0.4) reflects an overall performance that is somewhat unbalanced, with a strong emphasis on precision but at the cost of missing a

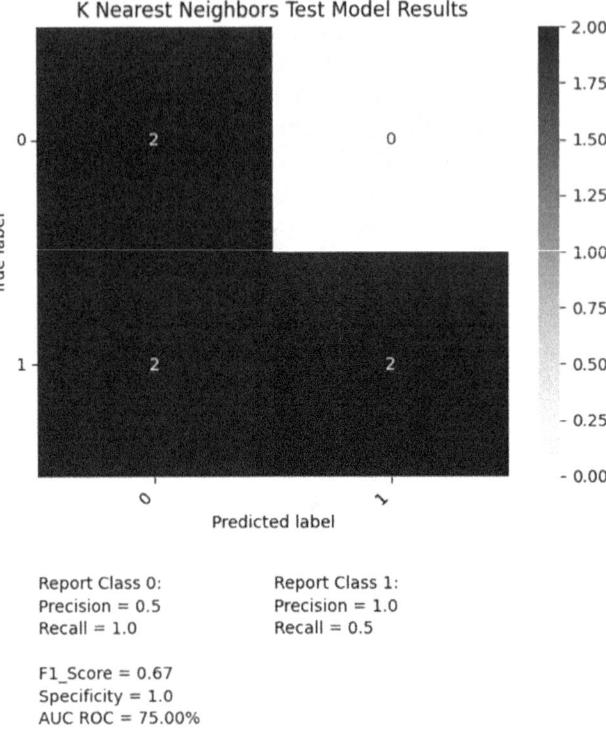

Report Class 0:
Precision = 0.5
Recall = 1.0

Report Class 1:
Precision = 1.0
Recall = 0.5

F1_Score = 0.67
Specificity = 1.0
AUC ROC = 75.00%

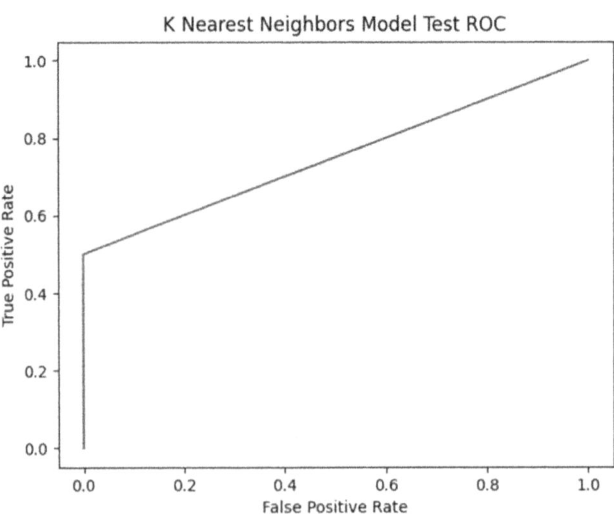

Figure 12(a) and (b): Confusion matrix and ROC for kNN test model for horizontal surface roughness

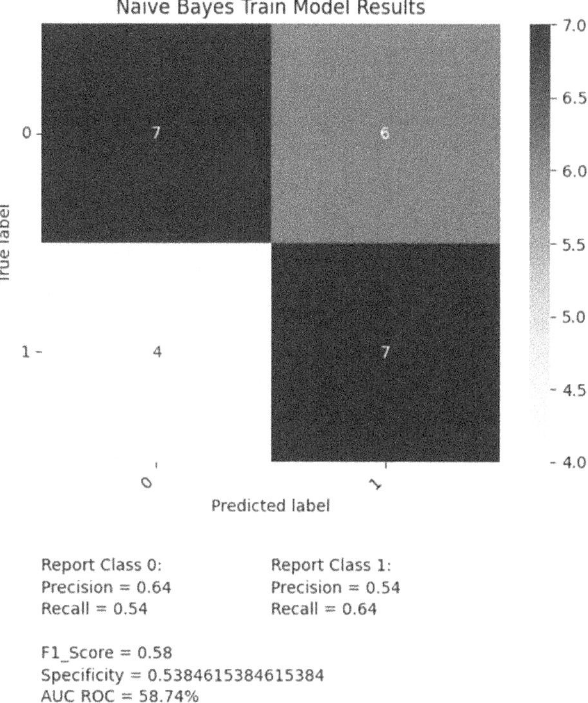

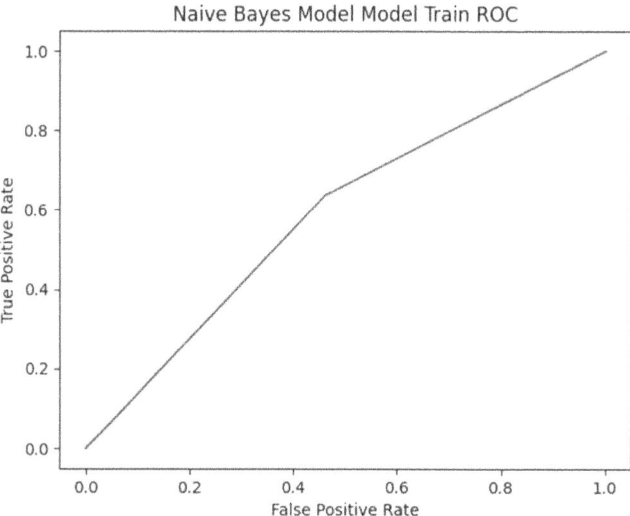

Figure 13(a) and (b): Confusion matrix and ROC for Naïve-Bayes train model for vertical surface roughness

significant portion of positive cases. Depending on the specific application, one may need to consider whether this trade-off is acceptable or if adjustments are needed to improve recall. The specificity is high (1), which means that the model can correctly identify all the negative cases in the train set. However, the AUC ROC is 62.5%, which suggests that the model can distinguish between the positive and negative cases fairly. Figure 14(a) and (b) depicts the Confusion matrix and ROC for Naïve-Bayes test model for vertical surface roughness, respectively.

Horizontal Surface Roughness

For class 0, the model has a high precision (0.81), implying that the model is generating a low number of false positives in comparison to true positives. The recall is also high (0.87), implying that it is correctly identifying many positive cases from the train set. For class 1, the model has good precision (0.75), meaning that when it predicts positive cases, it is correct about 75% of the time. It also has a moderate recall (0.67), indicating that it captures approximately 67% of the actual positive cases. The F1 score (0.71) reflects a well-balanced and solid overall performance, suggesting that your model is effective at both identifying positive cases and minimizing false positives. This is a promising performance for many classification tasks. The specificity is also high (0.8667), which means that the model can correctly identify most of the negative cases in the dataset. Additionally, the AUC ROC is 76.67%, which suggests that the model can distinguish between the positive and negative cases effectively. Figure 15(a) and (b) depicts the Confusion matrix and ROC for Naïve-Bayes train model for horizontal surface roughness, respectively.

For class 0, the model has a good precision (0.75), indicating that the model is generating a low number of false positives in comparison to true positives. The recall is perfect (1), indicating that the model is correctly identifying all the positive cases in the dataset. For class 1, the model has perfect precision (1.0), indicating that when it predicts positive cases, it is always correct. However, it has moderate recall (0.66), meaning that it identifies a significant amount of actual positive cases. The F1 score is 0.8, which signifies a balanced and solid overall performance. This indicates that your model is making accurate predictions with no false positives or false negatives, suggesting it is performing at the highest level of effectiveness for the given task. The specificity is also perfect (1), which means that the model can correctly identify all the negative cases in the test set. Additionally, the AUC ROC is 83.33%, which suggests that the model distinguishes the positive and negative cases effectively. Figure 16(a) and (b) depicts the Confusion matrix and ROC for Naïve-Bayes test model for horizontal surface roughness, respectively.

3.4 Logistic Regression Results

Vertical Surface Roughness

For class 0, the model has high precision (0.77), indicating that the model is

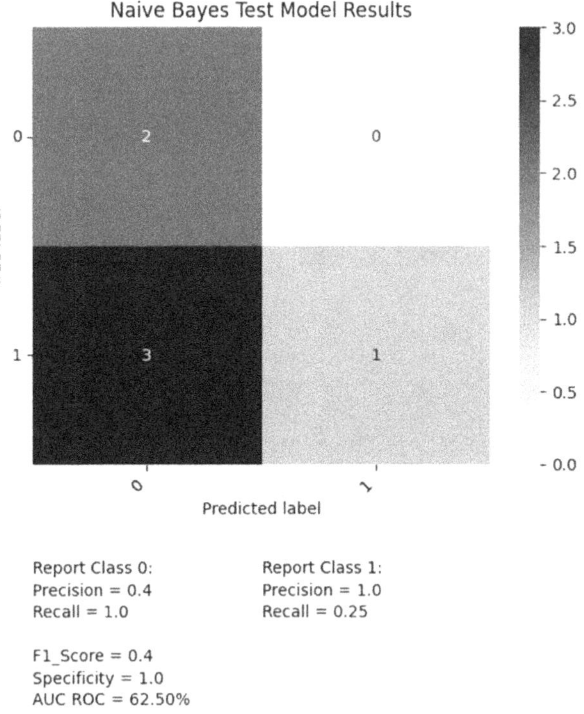

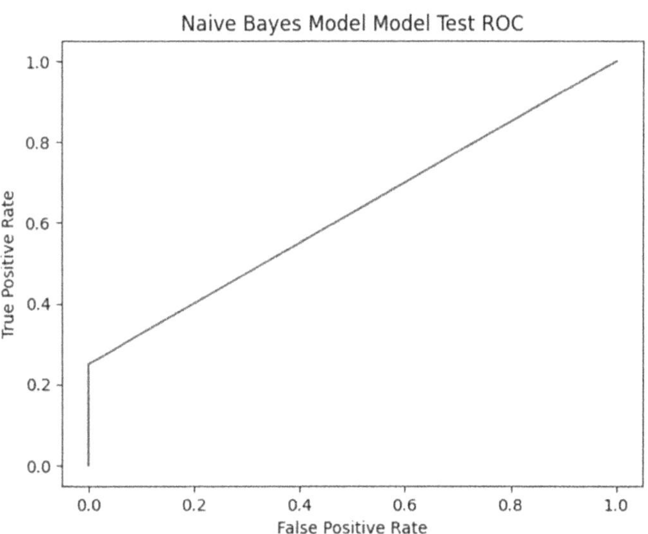

Figure 14(a) and (b): Confusion matrix and ROC for Naïve-Bayes test model for vertical surface roughness

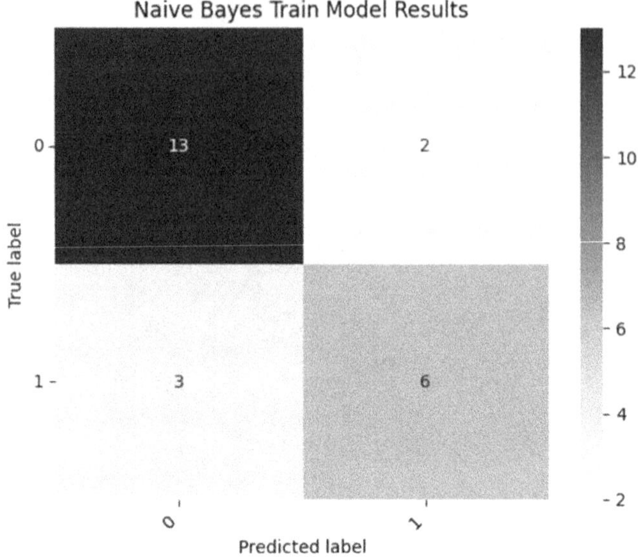

Report Class 0:
Precision = 0.81
Recall = 0.87

Report Class 1:
Precision = 0.75
Recall = 0.67

F1_Score = 0.71
Specificity = 0.8666666666666667
AUC ROC = 76.67%

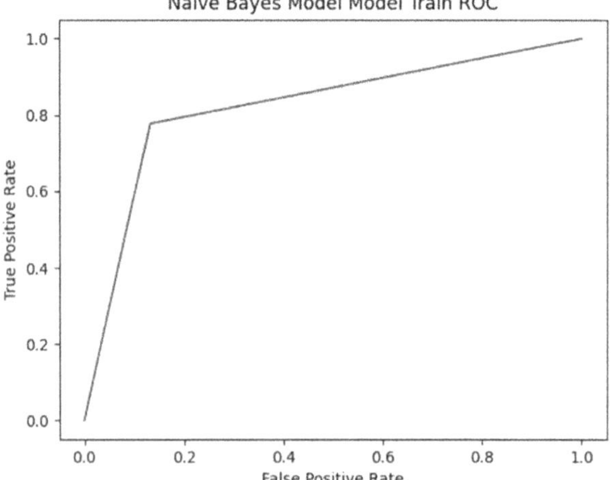

Figure 15(a) and (b): Confusion matrix and ROC for Naïve-Bayes train model for horizontal surface roughness

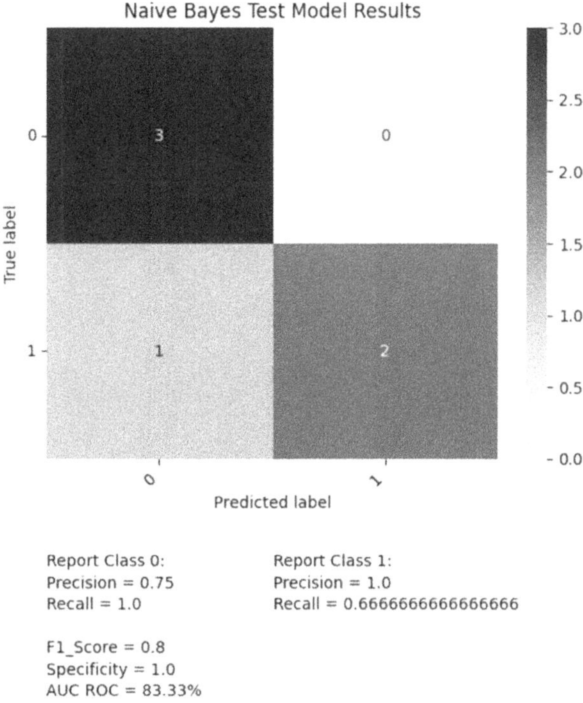

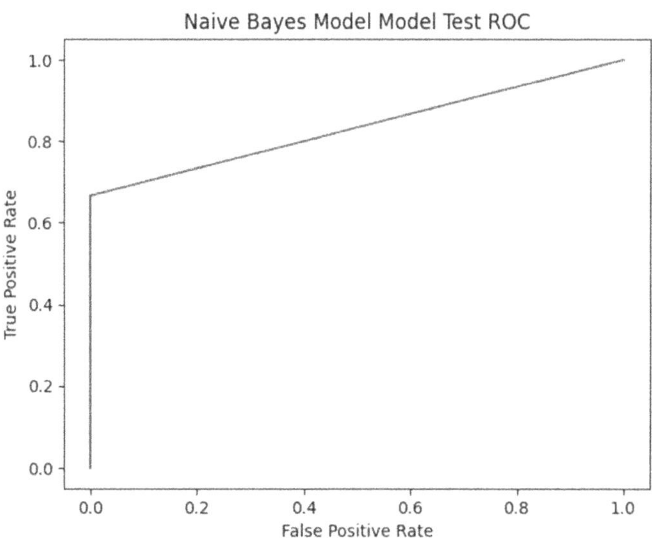

Figure 16(a) and (b): Confusion matrix and ROC for Naïve-Bayes test model for horizontal surface roughness

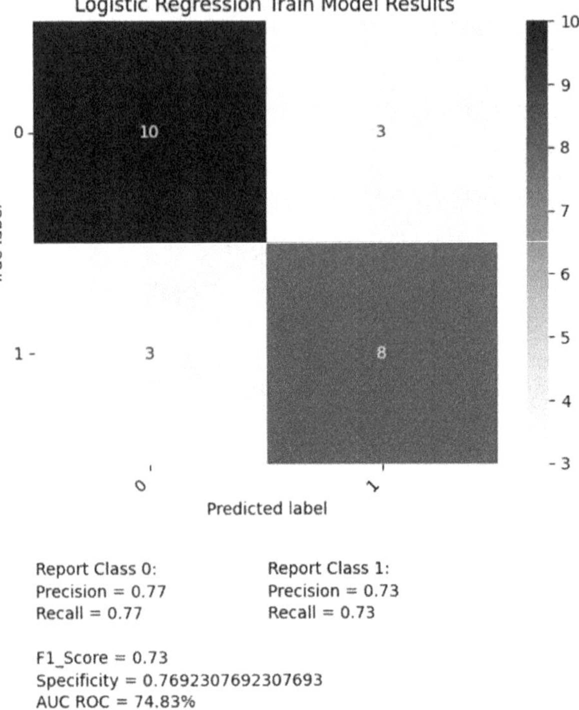

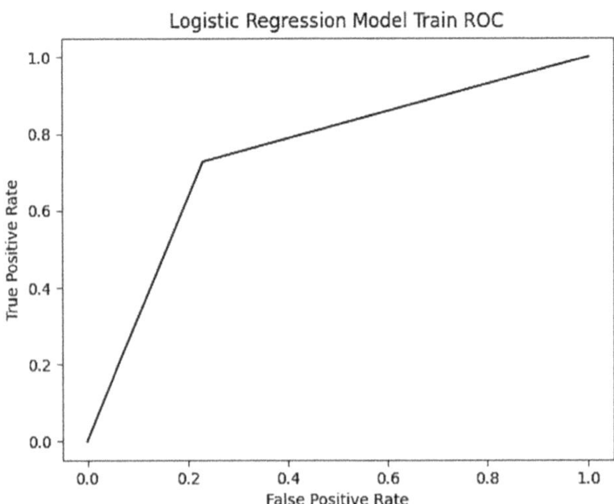

Figure 17(a) and (b): Confusion matrix and ROC for logistic regression train model for vertical surface roughness

generating a relatively low number of false positives compared to true positives. The recall is also relatively high (0.77), implying that the model is accurately identifying most of the positive cases in the training set. For class 1, the model has good precision (0.73), meaning that when it predicts positive cases, it is correct about 73% of the time. It also has a strong recall (0.73), indicating that it captures approximately 73% of the actual positive cases. The F1 score (0.73) reflects a balanced and solid overall performance, suggesting that your model effectively identifies positive cases while maintaining a reasonable level of precision. This is a respectable performance for many classification tasks. The specificity is also relatively high (0.7692), which means that the model can correctly identify most of the negative cases in the dataset. However, the AUC ROC is 74.83%, which suggests that the model cannot distinguish between the positive and negative cases perfectly. Figure 17(a) and (b) depicts the Confusion matrix and ROC for logistic regression train model for vertical surface roughness, respectively.

For class 0, the model has a relatively low precision (0.4), indicating that the model is generating a relatively high number of false positives compared to true positives. The recall is relatively high (1), indicating that the model is correctly identifying every positive case in the test set. For class 1, the model has perfect precision (1.0), meaning that when it predicts positive cases, it is always correct. However, it has a low recall (0.25), indicating that it captures only a quarter of the actual positive cases. Depending on the specific application, you may need to consider whether this trade-off is acceptable or if adjustments are needed to improve recall. This performance is suitable when minimizing false positives is of utmost importance, even if it means missing some positive cases. The F1 score (0.4) reflects an overall performance that is unbalanced, with a strong emphasis on precision but at the cost of missing a significant portion of positive cases. The specificity is perfect (1.0), which means that the model can correctly identify all the negative cases in the dataset. However, the AUC ROC is only 62.5%, which suggests that the model may cannot distinguish between the positive and negative cases effectively. Figure 18(a) and (b) depicts the Confusion matrix and ROC for logistic regression test model for vertical surface roughness, respectively.

Horizontal Surface Roughness

For class 0, the model has a very high precision (1.0), indicating that the model generates no number of false positives when compared to true positives. The recall is perfect (1.0), indicating that the model is correctly identifying all the positive cases in the train set. For class 1, the model has perfect precision (1.0), meaning that when it predicts positive cases, it is always correct. It also has a perfect recall (1.0), indicating that it captures 100% of the actual positive cases. The F1 score (1.0) reflects an outstanding overall performance with a strong balance between precision and recall. This suggests that your model is highly effective at both identifying positive cases and minimizing false positives, making it an excellent choice for the task at hand. The specificity is also perfect (1.0), which means that the model can correctly identify all the negative cases in the train set. The AUC

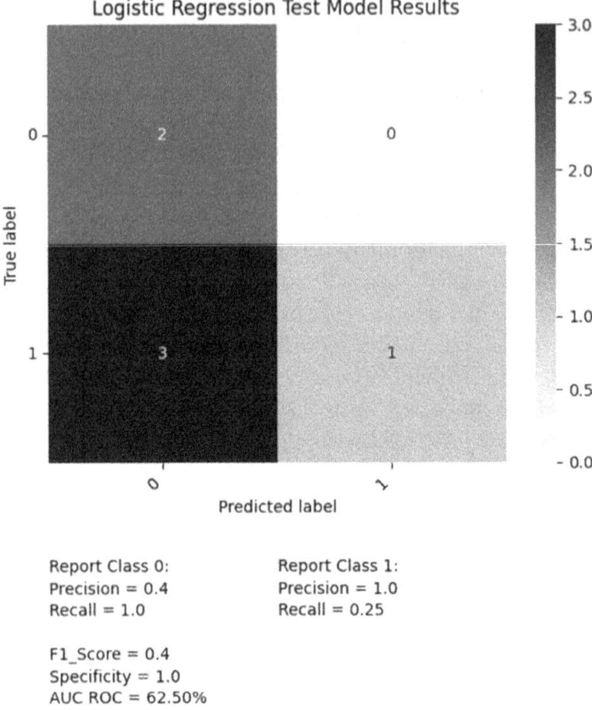

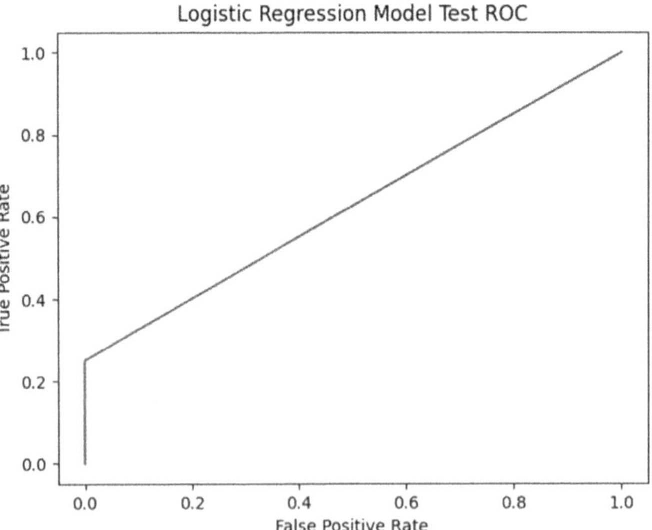

Figure 18(a) and (b): Confusion matrix and ROC for logistic regression test model for vertical surface roughness

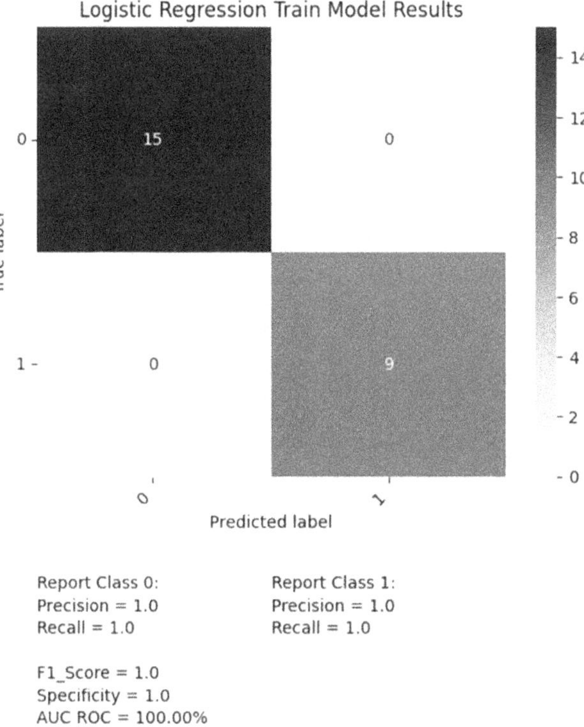

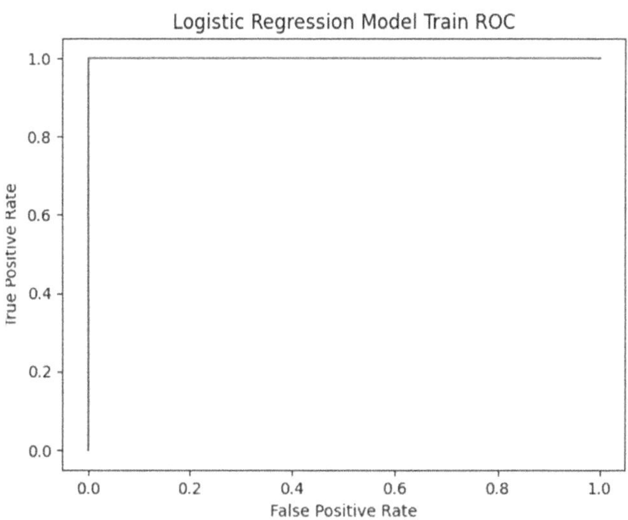

Figure 19(a) and (b): Confusion matrix and ROC for logistic regression train model for horizontal surface roughness

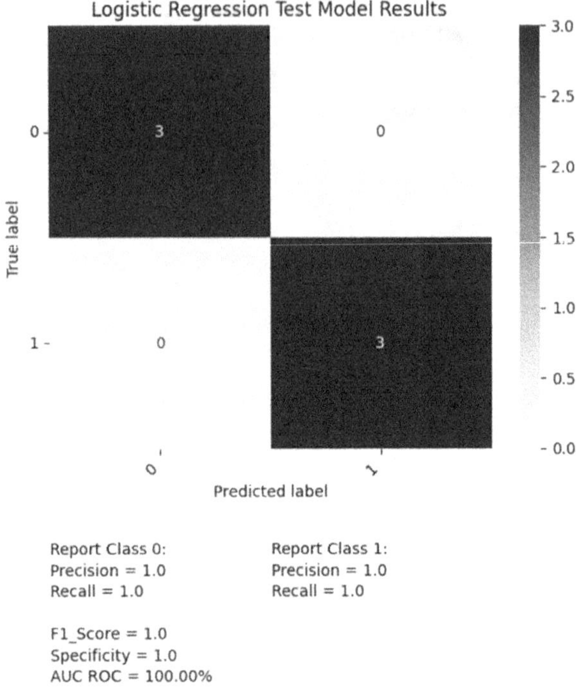

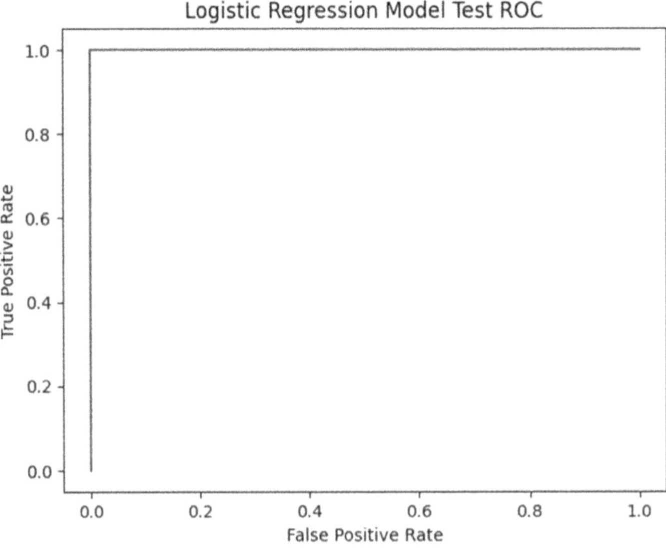

Figure 20(a) and (b): Confusion matrix and ROC for logistic regression test model for horizontal surface roughness

ROC is 100.0%, which suggests that the model can perfectly distinguish positive and negative cases. Figure 19(a) and (b) depicts the Confusion matrix and ROC for logistic regression train model for horizontal surface roughness, respectively.

For class 0, the model has perfect precision (1.0), indicating that the model is generating no false positives. The recall is also perfect (1.0), indicating that the model is correctly identifying the positive cases in the test set. For class 1, the model has perfect precision (1.0), meaning that when it predicts positive cases, it is always correct. It also has perfect recall (1.0), indicating that it correctly identifies all actual positive cases. The F1 score is also 1.0, which signifies an exceptional and flawless overall performance. This indicates that the model is making accurate predictions with no false positives or false negatives, suggesting it is performing at the highest level of effectiveness for the given task. The specificity is also perfect (1.0), which means that the model can correctly identify all the negative cases in the test set. The AUC ROC is 100%, which suggests that the model can perfectly classify the positive and negative cases. Figure 20(a) and (b) depicts the Confusion matrix and ROC for logistic regression test model for horizontal surface roughness, respectively.

4. Conclusion

The present study employed supervised machine learning classifiers namely, logistic regression, K-nearest neighbouring, and the Navies Bayes algorithm to predict surface roughness values of 3D printed parts. Based on the results of the study, layer height influenced horizontal surface roughness the most, and wall thickness influenced vertical surface roughness the most. Based on the values of the F1-score and AUC ROC, it can be inferred that the Navies Bayes algorithm and logistic regression outperformed in classifying horizontal direction surface roughness as well as vertical direction surface roughness. As a result, it can be concluded that supervised machine learning classifier algorithms are suitable for use in 3D printing manufacturing.

References

1. Medricky, R. and Fris, D. 2020. Analysis of the accuracy and the surface roughness of FDM/FFF technology and optimisation of process parameters. *Acta Polytechnica CTU Proceedings*, 7.1: 52–59.
2. Gómez-Gras, G., Pérez, P.M., Fabregas-Moreno, J. and Reyes-Pozo, G. 2021. Experimental study on the accuracy and surface quality of printed versus machined holes in PEI Ultem 9085 FDM specimens. *Materials Today: Proceedings*, 38: 107–112.
3. Thirugnanasambandham, K. and Kannan, K.V. 2020. Effect of printing parameters on the mechanical properties of parts fabricated by fused deposition modeling. *Materials Today: Proceedings*, 23: 220–225.

4. Chand, R., Sharma, V.S., Trehan, R., Gupta, M.K. and Sarikaya, M. 2020. Investigating the dimensional accuracy and surface roughness for 3D printed parts using a multi-jet printer. *Materials Today: Proceedings*, 23: 316–322.
5. Vogeler, F., Pee, B.V., Jansen, H. and Faes, M. 2018. Dimensional accuracy and surface texture of additive manufactured plastic inserts for injection moulding. *Materials Today: Proceedings*, 5.5: 11421–11428.
6. Yousaf, D. and Javdanierfani, K. 2021. The Effects of Build Orientation on the Printing Quality. Thesis in Mechanical Engineering, Halmstad.
7. Yang, Z. and Huang, Y. 2021. Improved classification algorithm based on genetic programming and its application in process monitoring of additive manufacturing. *Advances in Mechanical Engineering*, 13.1: 1687814021990937.
8. Qi, X., Chen, G., Li, Y., Cheng, X. and Li, C. 2020. Applying Neural-network-based machine learning to additive manufacturing: Current applications, challenges, and future perspectives. *Journal of Manufacturing Science and Engineering*, 143.2: 021010.
9. Equbal, A. 2021. Application of machine learning in fused deposition modelling: A review. *Additive Manufacturing and Rapid Prototyping*, 313–326.
10. Dehoff, R.R., Kirka, M.M., Sames, W.J. and Babu, S.S. 2018. Real-time quality control of metal powder bed fusion additive manufacturing using convolutional neural networks. *Scientific Reports*, 8(1): 1–8.
11. Hanon, M.M., Zsidai, L. and Ma, Q. 2020. Accuracy investigation of 3D printed PLA with various process parameters and different colours. *The International Journal of Advanced Manufacturing Technology*, 108.9-10: 2801–2816.
12. Selvam, A., Mayilswamy, S., Whenish, R., Naresh, K., Shanmugam, V. and Das, O. 2020. Multi-objective optimization and prediction of surface roughness and printing time in FFF printed ABS polymer. *The International Journal of Advanced Manufacturing Technology*, 108.9-10: 2969–2987.
13. Rizea, A.D., Anghel, D.C. and Iordache, D.M. 2018. Study of the deviation of shape for the parts obtained by additive manufacturing. *The International Journal of Advanced Manufacturing Technology*, 97.9-12: 4023–4032.
14. Karnati, S., Kumar, A. and De, A. 2021. Process parameter optimization of additive manufacturing using artificial neural networks. *Materials Today: Proceedings*, 43: 372–377.
15. Islam, M., Park, J.H. and Ahn, S.H. 2019. Machine learning-based predictive modeling of mechanical properties for additively manufactured polymers. *Additive Manufacturing*, 27: 190–200.

CHAPTER 11

Polymer Composite Impact Strength Estimation using K-Nearest Neighbouring Classification Algorithm

Dhruv A. Sawant[1*], Ashwini V. Jatti[2]

[1] Department of Mechanical Engineering, Symbiosis Institute of Technology, Symbiosis International (Deemed University), Pune, India

[2] Department of Instrumentation Engineering, Dr. D.Y. Patil Institute of Technology, Savitribai Phule Pune University, Pune, India

1. Introduction

Holding the top rank on the global market is the aim of manufacturers all over the world in today's technologically sophisticated industrial environment. In this, every conceivable branch of technology is attempted to be produced. The development of the entire prototype and its design laid the groundwork for the creation of the finished item. As the main raw materials for manufacturing currently, polymer matrix composite materials are taking the lead over fibre plastic, metals, and alloys. In light of the aforementioned, the objective of the current study was to create a novel polymer composite matrix material with superior mechanical properties that may be dependable and satisfy the growing demand from the manufacturing sectors on the global market. Due to the technological, mechanical, and physical characterization necessary for significant engineering applications, researchers, technicians, and scientists have discovered that hybrid composites with polymeric matrix materials saturated with reinforcers and particles in the form of molecules are of great interest [1]. Polymer composites have been used in many structural parts in the aerospace and automotive sectors for decades due to their lightweight

*Corresponding author: dhruv.sawant.btech2022@sitpune.edu.in

and outstanding mechanical qualities [2]. To improve the impact kinetics of the composite structures, filler material was included into polymer composite structures using hybridization procedures [3]. Over the past few decades, metal polymer composites, lightweight, high-performance materials for industrial applications, have gained popularity. These materials combine low density, high strength composites with extremely ductile and durable metals [4]. In order to improve the thermal and mechanical properties of acrylonitrile butadiene styrene (ABS) nanocomposites, polymeric carbon nitride nanosheets (NPCN) were created and modified [5]. Because ferroelectric polymers have a low energy efficiency due to significant intrinsic hysteretic losses, the best polymer matrix is the linear dielectric acrylonitrile butadiene styrene (ABS), which is inexpensive and has good film properties [6]. The notched impact strength and elastic moduli of polymers are the mechanical qualities that need to be improved [7]. The most popular coating method for metalizing common polymers like acrylonitrile-butadiene-styrene (ABS) is electrochemical plating [8]. This study aimed to examine the physical and mechanical characteristics of three different plastics—Polylactic acid (PLA), Acrylonitrile Butadiene Styrene (ABS), and Nylon, produced utilizing fused deposition modelling (FDM) and traditional injection modelling methods [9]. For PC/ABS and PBT/ABS type blends, the sustainable filler system made up of biocarbon (BC) particles and short basalt (BF) and carbon (CF) fibres were employed as reinforcement [10]. The most important and crucial requirements for creating effective EM interference (EMI) shielding polymer-based composite foams are low reflection and high absorption characteristics for electromagnetic (EM) waves [11]. The composites' thermal (DSC and TGA), morphological, and tensile characteristics were investigated [12]. The class of smart materials includes shape memory polymers (SMPs). When a stimulus, such as heat or PH, is applied, they have the capacity to change from a temporary shape back to its permanent shape [13].

The current study examines the impact strength of various ABS-Cu and ABS-Al composition variations. The paper is broken up into three sections: the first section covers the tools and procedures used to test the impact resistance of impact specimens; the second section covers the generalization of machine learning classification; the results; and discussions of the experimental data recorded; and the third section deals with the analysis of the experimental data of ABS-Cu and ABS-Al using machine learning classification.

2. Materials and Methods

There must be several compositions of the same composite material in order to attain the best mechanical attributes in terms of strength. This percentage by weight of each primary substance (ABS) uses surfactant (noninphinoethoxylate) and metallic powder (copper 99.9% pure). The compositions are shown in Figs. 1(a) and 1(b). In this instance, it is believed that the substance weighs 250 grams. The percent by the weight of copper should be maintained with a minimum 20%

Figure 1(a): ABS-Cu impact specimen

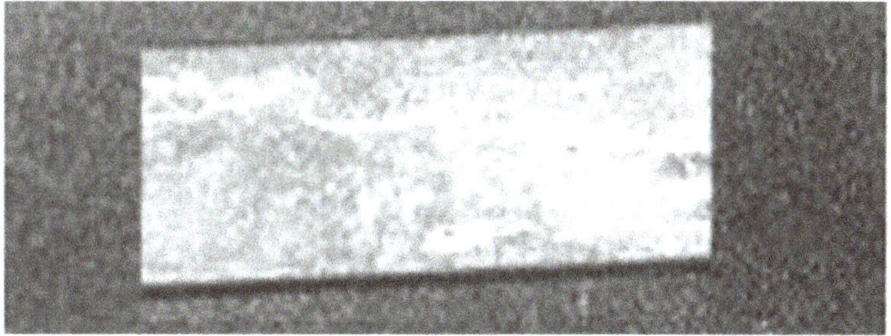

Figure 1(b): ABS-Al impact specimen

difference in order to precisely mould and enhance the mechanical properties of composite material. The test demonstrates a material's capacity to endure a sudden impact from a heavy load. It provides the amount of energy that the item must have absorbed to shatter at a high speed. From the standpoint of the designer, maintaining the quality and liability of the product is a crucial characteristic of the specimen or material. When designing the essential component of the product to improve its mechanical qualities, impact resistance is considered for numerous applications. Samples were chosen in accordance with ASTM D256 standard. The specimens had to have a good V-notch in accordance with standards; any flaws would have a negative impact on the impact test. The test was carried out by attaching the double anvil support device firmly using the four machine screws provided for the base plate to the impact tester's base. Then, the computer was set up to record data. The samples were then put into the fixture, and information was electronically recorded.

3. Machine Learning

In this work, a variety of machine learning classification algorithms for diverse material compositions (including copper, aluminum, ABS, and surfactant material) were used, along with sample impact testing.

The synthetic data generation procedure in Matlab was used to create sample data based on the experimental data in order to accurately analyze the impact strength using classification and to depict confusion matrices and AUC-ROC curves for a more in-depth study. The Classification algorithm is a Supervised Learning method that classifies fresh observations based on training data. In classification, a program uses the dataset or provided observations to learn how to classify new observations into different groups or classes. To divide the data into two distinct groups in the current study, the value above the average of the impact strength values is regarded to be 1, and the value below it is considered to be 0. The k-nearest neighbour (kNN) classification algorithm, which calculates the shortest Euclidean distance between the average value and each value of impact strength, was used in the current investigation.

1. The number of neighbours hyperparameter option determines how many nearest neighbours in the target dataset will be used to classify each value or point in the kNN classification method.
2. The distance metric, which calculates the separation between two places.
3. The distance weight (1/distance) determines whether the distance is equal or inverse.

The metrics module from the sklearn library of the Python language was used to plot the confusion matrix. To obtain accurate results for the data prediction, the dataset was divided into 80% training data and 20% random test data. The True positives (TP) and True negatives (TN) in the confusion matrix represent successfully predicted data, whereas the False positives (FP) and False negatives (FN) represent incorrectly predicted data or indicate a mistake in the data prediction. Only the values of k=1, 2, 3, and 5 were taken into consideration for the number of neighbours (k) in order to gather a wide variety of classification data. The AUC-ROC curve is a performance measure for classification problems at various threshold levels. ROC is a probability curve, and AUC stands for the level or measurement of separability. It demonstrates how well the model can distinguish between classes. The greater the AUC, the better the model is at categorizing 0 class as 0, and 1 class as 1.

4. Results and Discussions

The composition ABS-Cu with ABS% equal to 64.89, copper% equal to 29.92 and surfactant material percentage equal to 4.71 had the highest impact strength, which was 8.53, according to the graph in Fig. 2(a). This illustrates how, as copper content increases, impact strength decreases. The composition ABS-Al exhibited the maximum impact strength, measuring 6.75, with ABS content of 65.14 percent, aluminum content of 29.86 percent, and surfactant material content of 5.01 percent. This shows that the impact strength decreases as the amount of aluminum in the polymer matrix increases. Figures from ABS-Cu and ABS-Al composite impact testing are presented in Figs. 4(a), 4(b), and 4(c), and Figs. 5(a), 5(b), and 5(c), respectively.

Figure 2(a)'s heatmap illustrates the relationship between ABS, copper, surfactant, and the impact strength of an ABS-Cu composite material. According to the F-test results, ABS% composition significantly affects impact strength while surfactant material% and copper% has a comparatively less impact.

As shown in the F-test and heatmap correlation matrix for ABS-Al in Figs. 3(a) and 3(b), surfactant material has a substantial impact on the composition of ABS-Al, followed by ABS% and aluminum%.

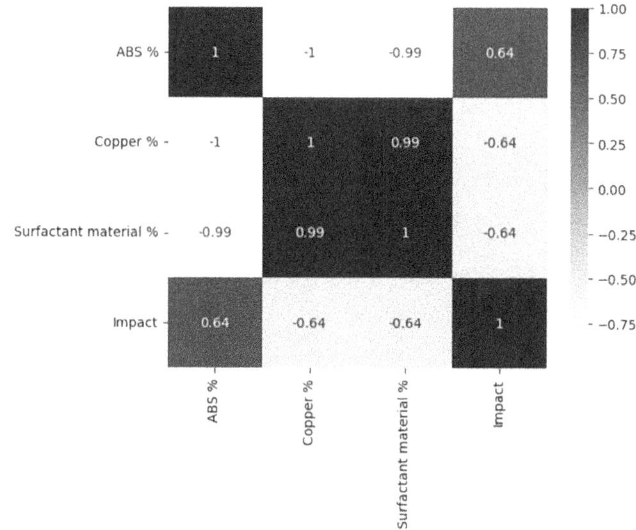

Figure 2(a): ABS-Cu composition heatmap

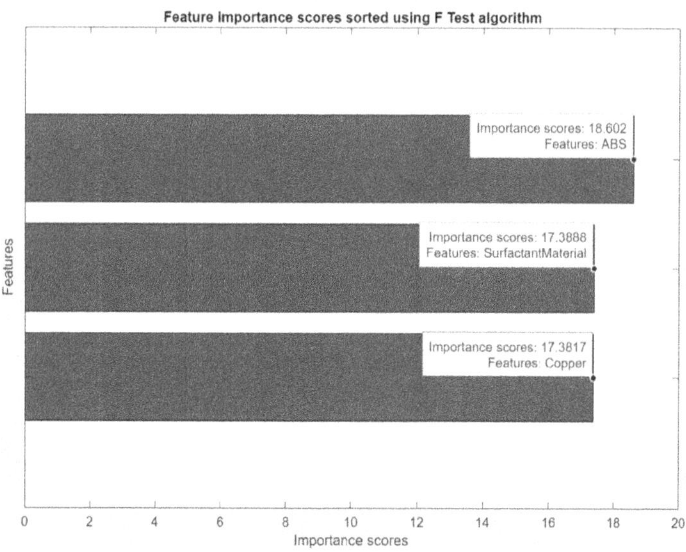

Figure 2(b): F-test for ABS-Cu feature selection

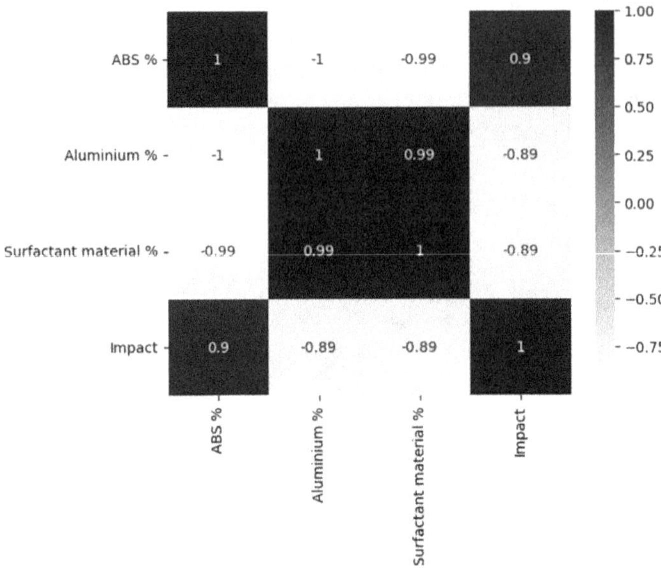

Figure 3(a): ABS-Al composition heatmap

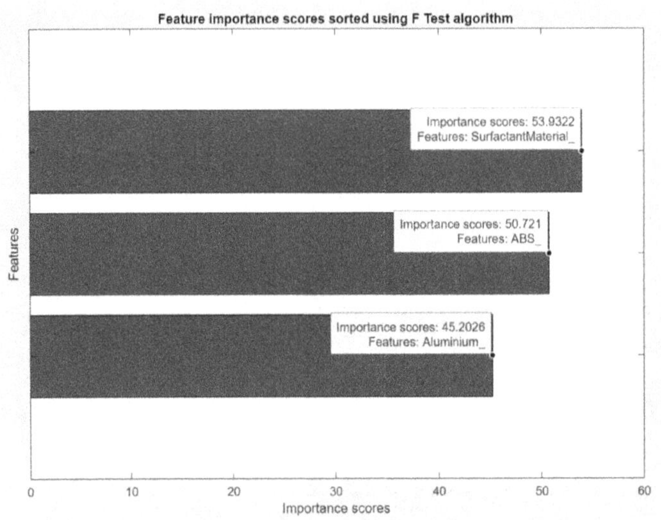

Figure 3(b): F-test for ABS-Al feature selection

4.1 Machine Learning Classification: ABS-Cu

For the current investigation, the k-nearest neighbour classification accuracy for impact strength of ABS-Cu composition for k=1, 2, 3 and 5 for the impact strength of ABS-Cu composition was 80%, 75%, 85%, and 95%, respectively.

Figure 4(a): ABS-Cu (Copper 60% + ABS 35%) after impact test

Figure 4(b): ABS-Cu (Copper 50% + ABS 44%) after impact test

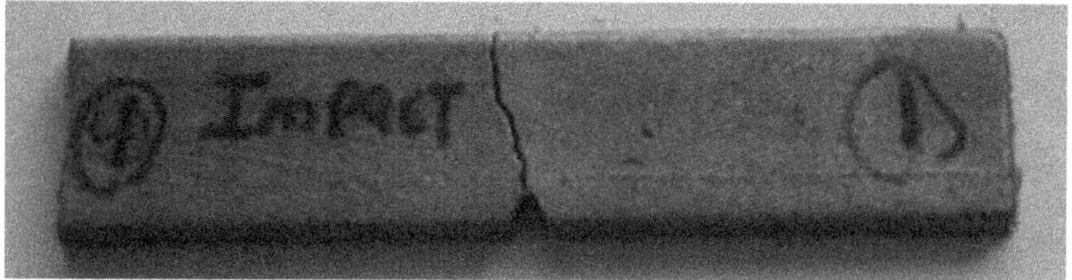

Figure 4(c): ABS-Cu (Copper 70% + ABS 23%) after impact test

Figure 5(a): ABS-Al (Aluminum 30% + ABS 35%) after impact test

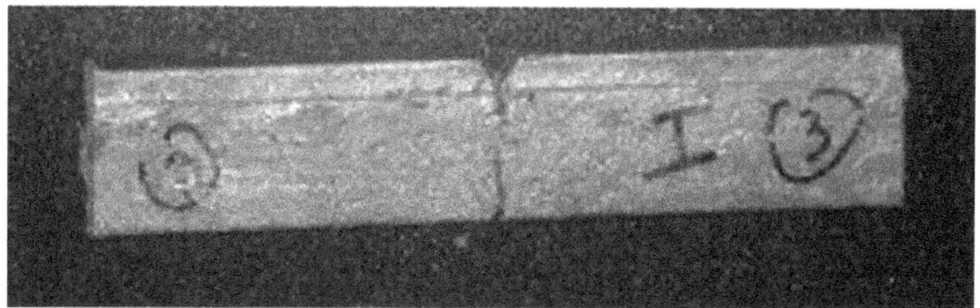

Figure 5(b): ABS-Al (Aluminum 50% + ABS 44%) after impact test

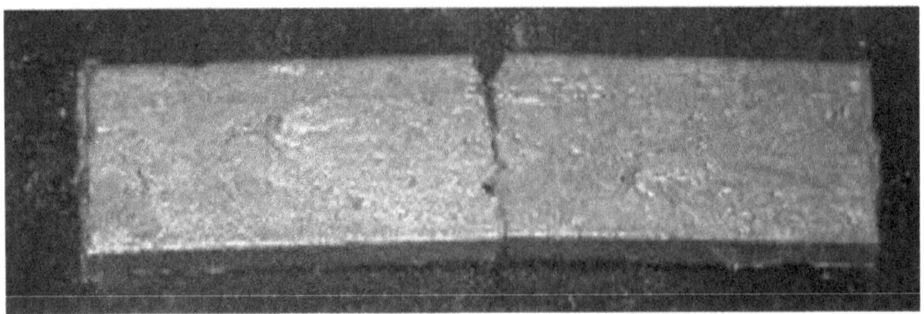

Figure 5(c): ABS-Al (Aluminum 70% + ABS 23%) after impact test

The Euclidean distance between the nearest and average values of impact strength is calculated using the k-nearest neighbour method. The values that were true positives (1) were considered as impact strength values above the average value which was 5.72 and true negatives (0) were considered below the average value. The confusion matrix is divided into 4 quadrants for study.

1. Top-left quadrant: True Negative
2. False Positive in the Top-Right Quadrant
3. False Negative in the Bottom Left Quadrant 3.
4. Genuine Positivity (lower-right quadrant).

The precision, recall, F1-score, and support are shown in the classification report in Table 1(a) for ABS-Cu. The impact strength prediction's precision and recall values are accurate because they are greater than 0.75, which indicates that the data was predicted with 75% accuracy or above. The AUC-ROC curve for ABS-Cu has a value of 0.8011, which is nearer 1, showing that the classification model correctly predicted the findings for impact strength.

4.2 Machine Learning Classification: ABS-Al

Similar to this, the data for the impact strength of ABS-Al composition was classified using k-nearest neighbours. The values above the average value of impact strength which was 5.29 were considered 1 and the values below the average values were considered 0, For k=1, 2, 3, and 5, respectively, the classification accuracy for the k-nearest neighbour classification for the impact strength of the ABS-Al composition was 95%, 90%, 95%, and 90%. The Euclidean distance between the nearest and average values of impact strength is calculated using the k-nearest neighbour method. The classification report for the impact strength of ABS-Al composition is presented in Table 1(b). It shows the precision, recall, f1-score, and support for the prediction accuracy of the classification model. The AUC-ROC curve for ABS-Cu shows a value of 0.8139, which is nearer to 1 as shown in Fig. 7(e), shows that the classification model accurately predicted the impact strength values.

Table 1(a): Classification report for ABS-Cu

Classifier	Precision	Recall	F1- score	Support
0	1.00	0.90	0.95	10
1	0.91	1.00	0.95	10
Accuracy	-	-	0.95	20
Macro Avg	0.95	0.95	0.95	20
Weighted Avg	0.95	0.95	0.95	20

Table 1(b): Classification report for ABS-Al

Classifier	Precision	Recall	F1-score	Support
0	0.88	1.00	0.93	7
1	1.00	0.92	0.96	13
Accuracy	-	-	0.95	20
Macro Avg	0.94	0.96	0.95	20
Weighted Avg	0.96	0.95	0.95	20

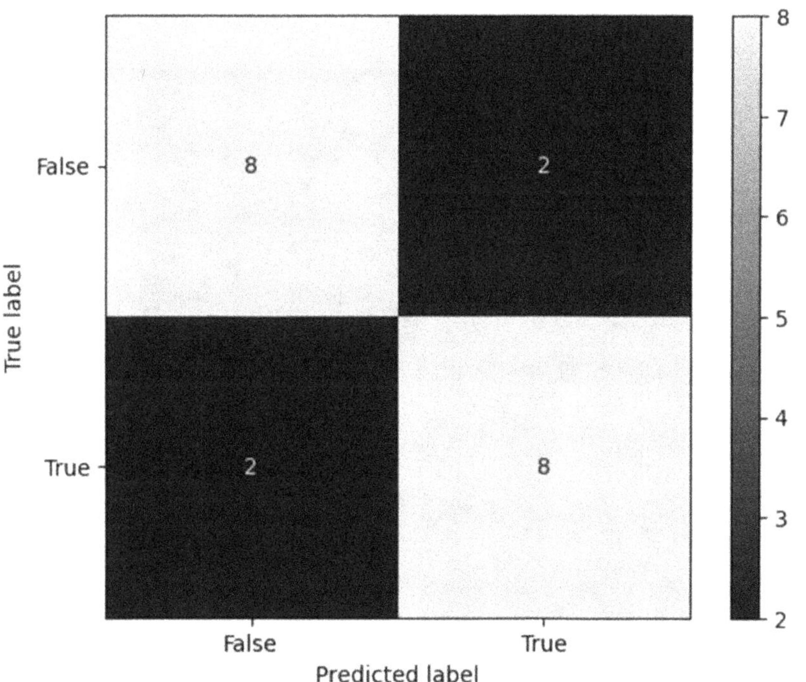

Figure 6(a): ABS-Cu confusion matrix for k=1

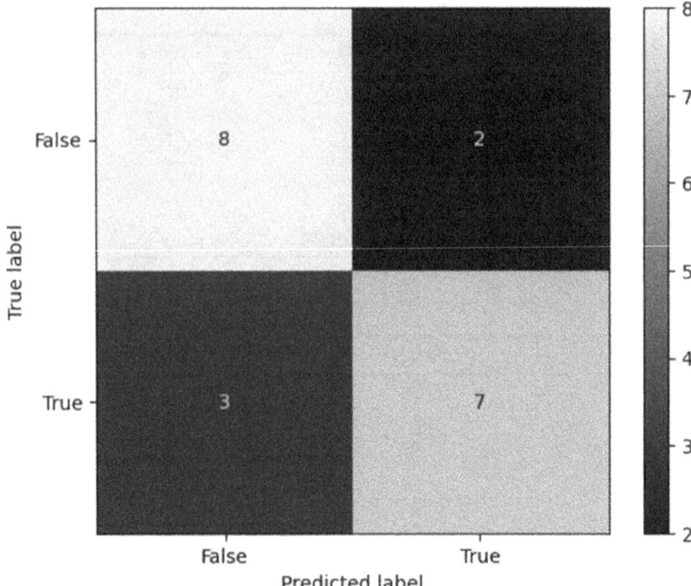

Figure 6(b): ABS-Cu confusion matrix for k=2

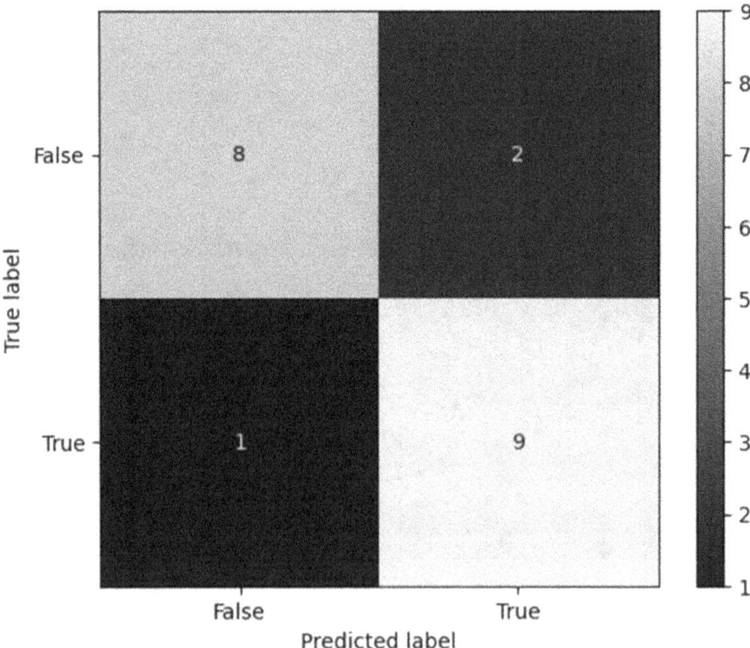

Figure 6(c): ABS-Cu confusion matrix for k=3

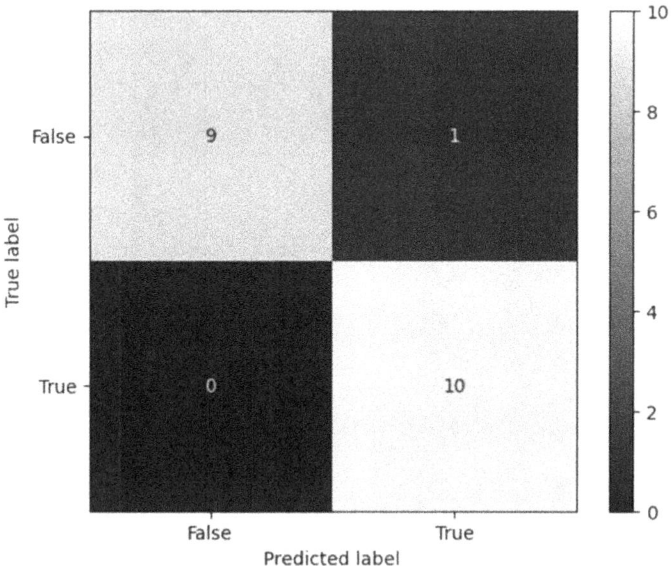

Figure 6(d): ABS-Cu confusion matrix for k=5

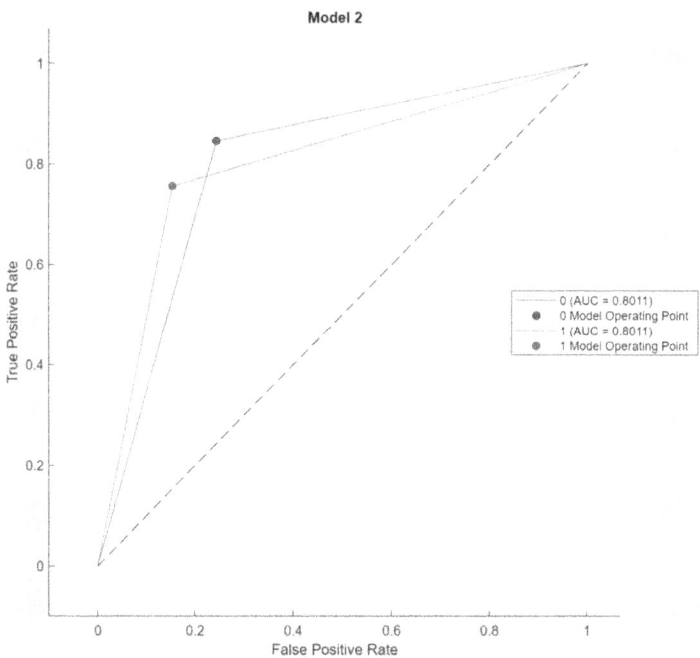

Figure 6(e): AUC-ROC curve for ABS-Cu

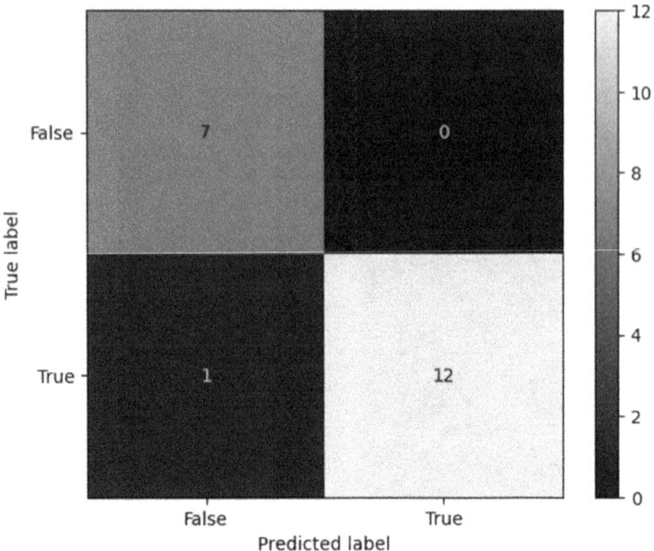

Figure 7(a): ABS-Al confusion matrix for k=1

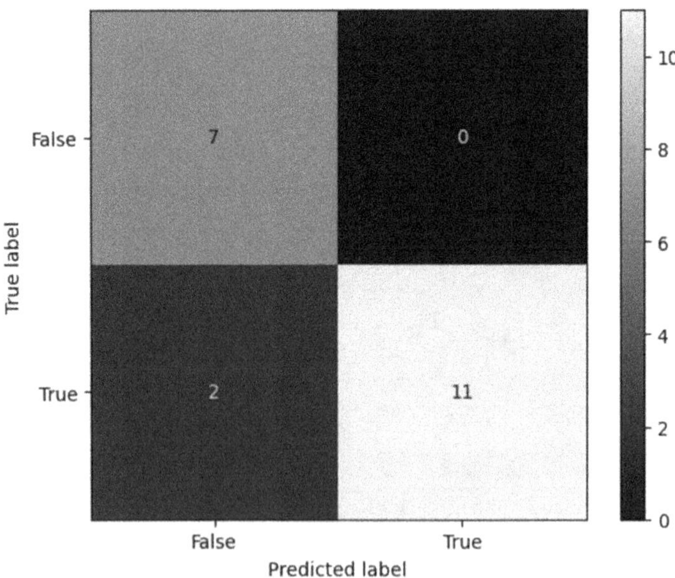

Figure 7(b): ABS-Al confusion matrix for k=2

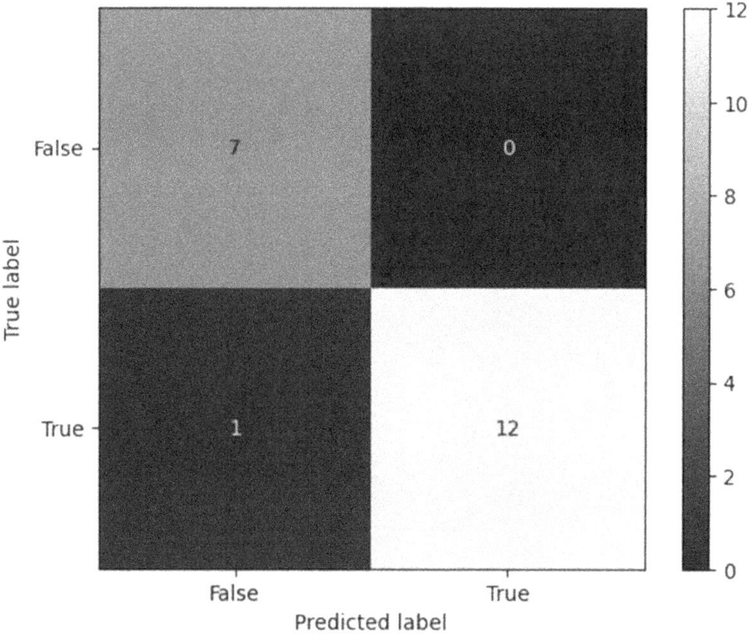

Figure 7(c): ABS-Al confusion matrix for k=3

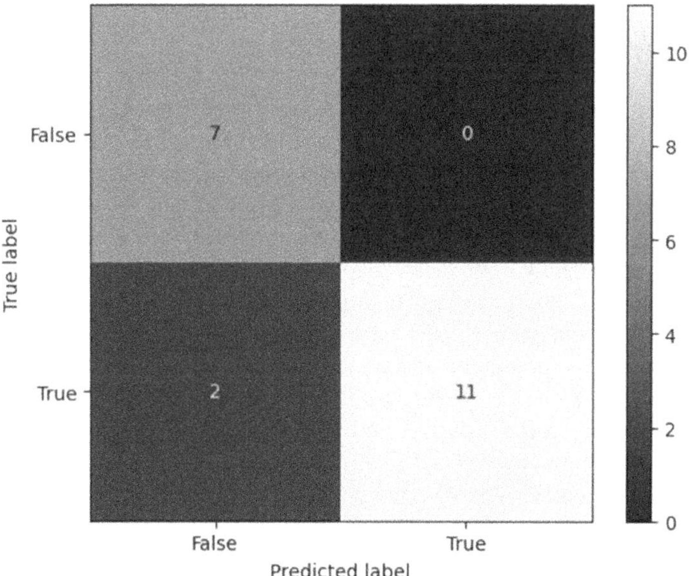

Figure 7(d): ABS-Al confusion matrix for k=5

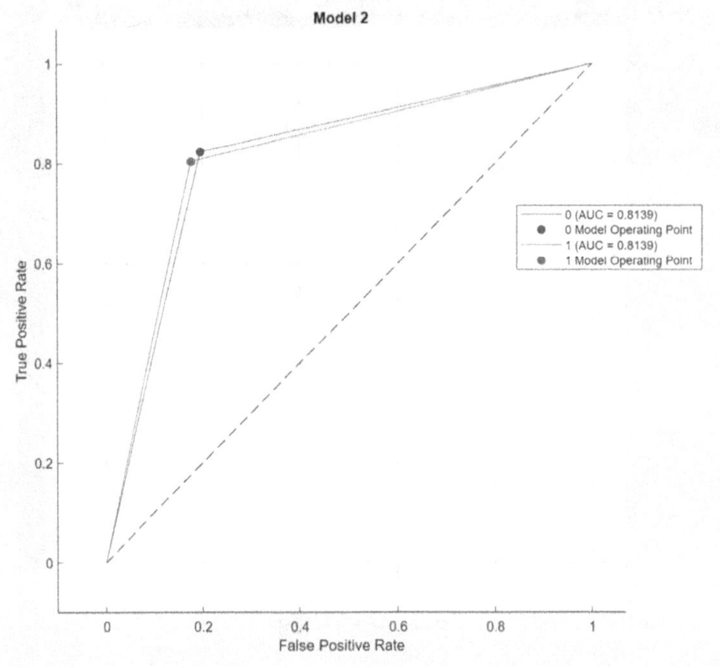

Figure 7(e): AUC-ROC curve for ABS-Al

5. Conclusion

The primary objective of this study was to investigate how modifications in the percentage compositions of ABS, copper, and aluminum in ABS-Cu and ABS-Al composite materials affected the impact strength of the corresponding composite materials. In ABS-Cu, the proportion impacts of the surfactant, copper, and ABS were roughly equal to 30%, while the percentage effects of the surfactant, aluminum, and ABS were roughly equivalent to 75%. Impact strength was predicted using machine learning classification, which involved plotting confusion matrices for true positives, true negatives, false positives, and false negatives with values that are near 100%. The prediction accuracy values for ABS-Cu for k=1, 2, 3, and 5 were 80%, 75%, 85%, and 95% respectively, and 95%, 90%, 95%, and 90% for ABS-Al composition for k=1, 2, 3, and 5 respectively. These numbers show that the prediction is correct. The classification report's precision, recall, and accuracy numbers are likewise closer to 1, showing that the impact strength forecast is quite accurate. The emphasis of this work will instead be on a careful examination of the mechanical characteristics and strength of reinforced carbon fibres.

References

1. Noor, H.A., Suha, K.S. and Muzher, T.M. 2022. Mechanical and physical characteristics of hybrid particles/fibers-polymer composites: A review. *Materials Today: Proceedings*, 62(1): 178-183, ISSN 2214–7853, https://doi.org/10.1016/j.matpr.2022.02.614.
2. Swagatika, M. and Punyapriya, M. 2022. Synthesis and characterization of CCTO0.5-BT0.5/epoxy hybrid ceramic polymer composite for electronic applications. *Materials Today: Proceedings*, 67(2): 372-376, ISSN 2214–7853, https://doi.org/10.1016/j.matpr.2022.07.244.
3. Megavannan, M., Thiyagu, M. and Pradeep, K.K. 2023. Review of the effects of low-velocity impact events on advanced fiber-reinforced polymer composite structures. *Materials Today: Proceedings*, ISSN 2214–7853, https://doi.org/10.1016/j.matpr.2023.04.255.
4. Deepak, K.P., Ahmed, A.H., David, N., Nikolaos, T., Mitchell, L.R., Vipin, K. et al. 2023. A novel additive manufacturing compression overmolding process for hybrid metal polymer composite structures. *Additive Manufacturing Letters*, 5: 100128, ISSN 2772–3690, https://doi.org/10.1016/j.addlet.2023.100128.
5. Elssa, G., Abhisha, M., Poornima, V.P., Henri, V., Soney, C.G. and Saithalavi, A. 2023. Polydopamine modified polymeric carbon nitride nanosheet based ABS nanocomposites for better thermal, frictional and mechanical performance. *Nano-Structures & Nano-Objects*, 35: 100987, ISSN 2352–507X, https://doi.org/10.1016/j.nanoso.2023.100987.
6. Cuilian, D., Shiqi, Y., Xinxuan, T., Zijin, L., Hang, L., Yan, Z. et al. 2022. The design and preparation of high-performance ABS-based dielectric composites via introducing core-shell polar polymers BaTiO3 nanoparticles. *Composites Part A: Applied Science and Manufacturing*, 163: 107214, ISSN 1359-835X, https://doi.org/10.1016/j.compositesa.2022.107214.
7. Jani, P., Carlos, B., Hany, A., Laura, S. and Martin, S. 2023. Compatibilized PC/ABS blends from solvent recycled PC and ABS polymers from electronic equipment waste. *Polymer Testing*, 120: 107969, ISSN 0142-9418, https://doi.org/10.1016/j.polymertesting.2023.107969.
8. Ruslan, M., Ran, T. and Gilles, L. 2023. Greener electrochemical plating of ABS polymer with unprecedented adhesion via hierarchical micro-nanotexturing. *Journal of Materials Research and Technology*, 24: 3575–3587, ISSN 2238–7854, https://doi.org/10.1016/j.jmrt.2023.04.001.
9. Venkatesh, R., Britto, J.J.J., Amudhan, K., Anbumalar, V., Prabhakaran, R. and Thiyanesh Sakhi, R. 2023. Experimental investigation of mechanical properties on CF reinforced PLA, ABS and Nylon composite part. *Materials Today: Proceedings*, 76(4): 647–653, ISSN 2214–7853, https://doi.org/10.1016/j.matpr.2022.12.091.
10. Jacek, A., Anna, D., Adam, P., Aminul, I. and Marek, S. 2023. Biocarbon-based sustainable reinforcing system for technical polymers. The structure-properties correlation between polycarbonate (PC) and polybutylene terephthalate (PBT)-based blends containing acrylonitrile-butadiene-styrene (ABS). *Sustainable Materials and Technologies*, 36: ISSN 2214–9937, https://doi.org/10.1016/j.susmat.2023.e00612.
11. Hai, F., Yu'an, B., Shuqian, D., Hongfu, Z. and Wei, G. 2023. Structure design of multi-layered ABS/CNTs composite foams for EMI shielding application with low reflection and high absorption characteristics. *Applied Surface Science*, 624: 157168, ISSN 0169-4332, https://doi.org/10.1016/j.apsusc.2023.157168.

12. Ramanjaneyulu, B., Venkatachalapathi, N. and Prasanthi, G. 2021. Thermal and mechanical properties of PLA/ABS/TCS polymer blend composites. *J. Inst. Eng. India Ser. C*, 102: 799–806. https://doi.org/10.1007/s40032-021-00687-7
13. Merve, U. and Melih, S.C. 2023. Evaluation of the bio-based materials utilization in shape memory polymer composites production. *European Polymer Journal*, 195: 112196, ISSN 0014–3057, https://doi.org/10.1016/j.eurpolymj.2023.112196.

Index

O

P

S

For Product Safety Concerns and Information please contact our EU representative GPSR@taylorandfrancis.com
Taylor & Francis Verlag GmbH, Kaufingerstraße 24, 80331 München, Germany

www.ingramcontent.com/pod-product-compliance
Lightning Source LLC
LaVergne TN
LVHW010601110826
845149LV00003B/730